Also by Paul Davies

The Physics of Time Asymmetry
Space and Time in the Modern Universe
The Runaway Universe
The Forces of Nature
Other Worlds
The Search for Gravity Waves
The Edge of Infinity
The Accidental Universe
Quantum Fields in Curved Space (with N. D. Birrell)
God and the New Physics
Superforce
Quantum Mechanics
The Ghost in the Atom (with J. R. Brown)
The Cosmic Blueprint
Fireball
Superstrings (with J. R. Brown)
The New Physics (editor)
The Mind of God
The Last Three Minutes
How to Build a Time Machine
Are We Alone?
About Time
The Fifth Miracle
The Big Questions (with Phillip Adams)
More Big Questions (with Phillip Adams)
Cosmic Jackpot

Also by John Gribbin

In Search of Schrödinger's Cat
In Search of the Big Bang
Schrödinger's Kittens and the Search for Reality
The Search for Superstrings, Symmetry, and the Theory of Everything
The Scientists
Deep Simplicity

THE MATTER MYTH

Dramatic Discoveries That Challenge
Our Understanding of Physical Reality

PAUL DAVIES
AND
JOHN GRIBBIN

Simon & Schuster Paperbacks
New York   London   Toronto   Sydney

Simon & Schuster Paperbacks
A Division of Simon & Schuster, Inc.
1230 Avenue of the Americas
New York, NY 10020

This Simon & Schuster trade paperback edition October 2007

SIMON & SCHUSTER PAPERBACKS and colophon
are registered trademarks of Simon & Schuster, Inc.

For information about special discounts for bulk purchases,
please contact Simon & Schuster Special Sales at
1-800-456-6798 or business@simonandschuster.com

Manufactured in the United States of America
1 3 5 7 9 10 8 6 4 2
The Library of Congress has cataloged
the Touchstone edition as follows:
Davies, P. C. W.
The matter myth: dramatic discoveries that challenge our
understanding of physical reality / Paul Davies and John Gribbin.
p.   cm.
"A Touchstone book."
Includes bibliographical references and index.
1. Reality.   2. Physics—Philosophy.   I. Gribbin, John R.
II. Title.
QC6.4.R42D38   1992
530—dc20   91-39522
CIP

ISBN-13: 978-0-7432-9091-3

Contents

Preface

The word "revolution" is rather overworked in science. Nevertheless, even those people with merely a casual interest in matters scientific will be aware that some truly revolutionary changes are currently taking place. We refer not so much to the specific discoveries that are happening all the time, nor to the many wonderful advances in technology. True, these changes are revolutionary enough in themselves. There is, however, a far more profound transformation taking place in the underlying science itself—in the way that scientists view their world.

The philosopher Thomas Kuhn has argued that scientists build their conception of reality around certain specific "paradigms." A paradigm is not a theory as such, but a framework of thought—a conceptual scheme—around which the data of experiment and observation are organized. From time to time in the history of ideas, a shift occurs in the basic paradigm. When this happens, not only do scientific theories change, but the scientists' conception of the world changes as well. That is what is happening now.

It has, unfortunately, become something of a cliché to claim that we are in the midst of such a paradigm shift. But such claims are usually based on only small portions of the truth. Many people will be aware that some strange and challenging notions have been surfacing in recent years: black holes, wormholes, quantum "ghostliness," chaos, "thinking" computers, to name but a few. These are, however, just the tip of a huge iceberg. In fact, as we approach the end of the twentieth century, science is throwing off the shackles of three centuries of thought in which a particular paradigm—called "mechanism"—has dominated the world view of scientists. In its simplest terms, mechanism is the belief that the physical Universe is nothing but a collection of material particles in interaction, a gigantic purposeless machine, of which the human body and brain are unimportant and insignificant parts. Mechanism, and the related philosophy of materialism, can be traced back to Ancient Greece; but their modern origins lie with Isaac Newton and his seventeenth-century contemporaries. It was Newton who gave us the laws of mechanics, and opened the way for the claim that all physical systems, all events, can be regarded as part of a vast mechanistic process. And it is this myth of materialism that is being laid as we move into the twenty-first century.

The movement toward a "postmechanistic" paradigm, a paradigm suitable for twenty-first-century science, is taking place across a broad front: in cosmology, in the chemistry of self-organizing systems, in the new physics of chaos, in quantum mechanics and particle physics, in the information sciences and (more reluctantly) at the interface of biology with physics. In all these areas scientists have found it fruitful, or even essential, to regard the portion of the Universe they are studying in entirely new terms, terms that bear little relation to the old ideas of materialism and the cosmic machine. This monumental paradigm shift is bringing with it a new perspective on human beings and their role in the great drama of nature.

Physicist Joseph Ford has described the materialistic, mechanistic paradigm as one of the "founding myths" of classical science. Myths, of course, are not literal expressions of truth. Are we to suppose, then, that the immense progress made in science during the past three hundred years is rooted in a complete misconception about the nature of nature? No. This would be to misunderstand the role of scientific paradigms. A particular paradigm is neither right nor wrong, but merely reflects a perspective, an aspect of reality that may prove more or less fruitful depending on circumstance—just as a myth, although not literally true, may contain allegorical insights that prove more or less fruitful depending on circumstances. In the event, the mechanistic paradigm proved so successful that there has been an almost universal tendency to identify it with reality, to see it not as a facet of the truth but as the whole truth. Now, increasing numbers of scientists are coming to recognize the limitations of the materialistic view of nature and to appreciate that there is more to the world than cogs in a gigantic machine.

In this book we explore these exciting and challenging changes and discuss their relevance to all of us, not just the scientists. In relating this story, we have to travel deep into scientific territory, but we have endeavored to keep the discussion as simple as possible, and in particular we have eschewed the use of mathematics entirely, even though some of the new concepts are fully meaningful only in mathematical language. Our intention is to provide a glimpse of the new Universe that is emerging. It is a picture still tantalizingly incomplete, yet compelling enough from what can already be discerned. We have no doubt that the revolution which we are immensely privileged and fortunate to be witnessing at firsthand will forever alter humankind's view of the Universe.

Paul Davies
John Gribbin
February 1991

1 The Death of Materialism

In daily life we are aware that some things change while others do not. We all grow older, and perhaps wiser, but the "we" to which these alterations happen apparently remains the same. Each day brings new events on Earth, but the Sun and stars seem unchanged. To what extent, though, are these merely human perceptions, limited by our senses?

In Ancient Greece, there raged a great debate about the nature of change. Some philosophers, such as Heraclitus, maintained that *everything* is in a state of flux; nothing escapes change of some sort. On the other hand, Parmenides argued that everything is what it is, so that it cannot become what it is not. Thus, change was incompatible with being, so that only the permanent aspects of the world could be considered truly real.

In the fifth century B.C. an ingenious escape from this dilemma was proposed by Democritus. He hypothesized that all matter is made up of tiny indestructible units, which he called atoms. The atoms themselves remained unchanging, having fixed properties such as size and shape, but they could move about in space and combine in various ways, so that the macroscopic bodies which they constitute might seem to alter. In this way, permanence and flux could be reconciled; all change in the world was attributed simply to the rearrangement of atoms in the void. Thus began the doctrine of materialism.

For centuries, materialism had to compete with other ideas: for example, with the belief that matter possessed magical or active qualities, or could be infused with vitalistic potency or occult forces. These mystical images faded with the rise of modern science. A key event in this regard occurred in the year 1687 with the publication of Isaac Newton's masterpiece, the *Principia*. It was in this work that Newton stated his famous laws of motion. Like the Greek Atomists before him, Newton treated matter as passive and inert. Indeed, *inertia* played a central role in his theory of the world. If a material body is at rest, then according to Newton's laws it will remain forever at rest unless acted upon by an external force. Similarly, if the body is moving, it will continue to move with the same speed and in the same direction unless a force acts to change it. Thus matter is entirely passive.

Newton's own words in this respect say it all. Matter consists of "solid, massy, impenetrable, movable particles." For Newton and his contemporaries, no essential distinction existed between the properties of everyday material objects and the elementary constituent particles that supposedly made up their substance, save in respect of the impenetrable quality of the latter.

The age of the machine

Newton's view of matter as inert substance shaped and formed by external forces became deeply ingrained in Western culture. It was to be embraced wholeheartedly during the Industrial Revolution, which brought immense power and wealth. In eighteenth- and nineteenth-century Europe, the forces of nature were being tamed, or harnessed for productive purposes. With steam and iron came locomotives and huge ships, and the power literally to alter the face of the Earth. And with these advances came a passion for possessing, in one form or another, large quantities of matter. Wealth was measured in acres of land, or in tons of coal, or gold or other commodities.

The Industrial Revolution was a time of fantastic confidence—the triumph of materialism. The confidence of the engineers did not rest merely upon the success of blind trial and error. There was an impressive body of knowledge and an understanding of the principles that underpinned the new machine age. These principles had been established by Newton two hundred years before, and elaborated by many others afterward.

At the time of the publication of the *Principia* the most sophisticated machines were clocks, and Newton's image of the working of nature as an elaborate clockwork struck a deep chord. The clock epitomized order, harmony and mathematical precision, ideas that fitted in well with the prevailing theology. Gone were the ancient notions of the cosmos as a living organism, imbued with mystical purposes. Newtonian mechanics had established a clear connection between cause and effect, and the mechanistic account required that matter move in accordance with strict mathematical laws. There was no room for mysterious active qualities. Indeed, the one realm that had retained overtones of magic and mystery—the heavens—provided the most successful application of Newton's mechanics. By combining his laws of motion with his law of gravitation, Newton was himself able to give a convincing account of the period of the Moon, and the orbits of planets and comets.

It is hard to overstate the impact that these physical images have had in shaping our world view. The doctrine that the physical Universe consists of inert matter locked into a sort of gigantic deterministic clockwork has penetrated all branches of human inquiry. Materialism dominates biology, for example. Living organisms are regarded as nothing more than complicated collections of particles, each being blindly pulled and pushed by its neighbors. Richard Dawkins, an eloquent champion of biological materialism, describes human beings (and other living entities) as "gene machines." Thus, organisms are treated as automatons. Such ideas have even influenced psychology. The behaviorist school treats all human activity in terms of a type of Newtonian dynamical system, in which the mind plays a passive (or inert) role and responds in an ultimately deterministic way to external forces or stimuli.

There is no doubt that the Newtonian world view, with its doctrine of materialism and the clockwork Universe, has contributed immensely to the advance of science by providing a highly intuitive framework within which to study a wide range of phenomena. But there is equally no doubt that it has also contributed in large part to alienating human beings from the Universe they inhabit. Donald Mackay, an expert on how the brain functions as a communications system, writes of "the disease of machine-mindedness." He points out that "In our age, when people look for explanations, the tendency more and more is to conceive of any and every situation that we are trying to understand by analogy with a machine." When extended into the domain of human affairs, such as politics or economics, machine-mindedness leads to demoralization and depersonalization. People feel a sense of helplessness; they are merely "cogs" in a machine that will lumber on regardless of their feelings or actions. Many people have rejected scientific values because they regard materialism as a sterile and bleak philosophy, which reduces human beings to automatons and leaves no room for free will or creativity. These people can take heart: materialism is dead.

New physics for a new social order

It is fitting that physics—the science that gave rise to materialism—should also signal the demise of materialism. During this century the new physics has blown apart the central tenets of materialist doctrine in a sequence of stunning developments. First came the theory of relativity, which demolished Newton's assumptions about space and time—assumptions that still hold sway in our everyday "common-sense" view of the world. The very arena in which the clockwork Universe acted out its drama was now exposed as subject to shifting and warping. Then came the quantum theory, which totally transformed our image of matter. The old assumption that the microscopic world of atoms was simply a scaled-down version of the everyday world had to be abandoned. Newton's deterministic machine was replaced by a shadowy and paradoxical conjunction of waves and particles, governed by the laws of chance rather than the rigid rules of causality. An extension of the quantum theory, known as quantum field theory, goes beyond even this; it paints a picture in which solid matter dissolves away, to be replaced by weird excitations and vibrations of invisible field energy. In this theory, little distinction remains between material substance and apparently empty space, which itself seethes with ephemeral quantum activity. The culmination of these ideas is the so-called superstring theory, which seeks to unite space, time and matter, and to build all of them from the vibrations of submicroscopic loops of invisible string inhabiting a ten-dimensional imaginary universe.

Quantum physics undermines materialism because it reveals that matter has far less "substance" than we might believe. But another development goes farther by demolishing

Newton's image of matter as inert lumps. This development is the theory of chaos, which has recently gained widespread attention. In fact, chaos is only part of a massive revolution in the way scientists now think about dynamical systems. It has been discovered that so-called nonlinear effects can cause matter to behave in seemingly miraculous ways, such as becoming "self-organizing" and developing patterns and structures spontaneously. Chaos as such is a special case of this; it occurs in nonlinear systems which become unstable and change in random and totally unpredictable ways. Thus the rigid determinism of Newton's clockwork Universe evaporates, to be replaced by a world in which the future is open, in which matter escapes its lumpen limitations and acquires an element of creativity.

In the coming chapters we shall be looking carefully at all these exhilarating developments and at the new world view which is emerging as a result. We shall see that matter as such has been demoted from its central role, to be replaced by concepts such as organization, complexity and information. This is already reshaping our social priorities. Consider, for example, the information technology revolution. The writer George Gilder has pointed out that the cost of the physical materials needed to make a silicon microchip is trifling, and he sees tomorrow's profits flowing to those countries and companies that can best market *information* and *organizational strategies*. The contrast with the matter-based wealth of the first Industrial Revolution could not be more striking:

> Today, the ascendant nations and corporations are masters not of land and material resources but of ideas and technologies. . . . The global network of telecommunications can carry more valuable goods than all the world's

supertankers. Wealth comes not to the rulers of slave labour but to the liberators of human creativity, not to the conquerors of land but to the emancipators of mind.

In this "overthrow of matter," writes Gilder, "the powers of mind are everywhere ascendant over the brute force of things," transforming "a material world composed of blank and inert particles to a radiant domain rich with sparks of informative energy."

No country more starkly faces the challenge of this transformation than Australia. For most of its history, the Australian economy has been dominated by the export of commodities such as coal, uranium and wool. For reasons of history and geography, very little manufacturing industry has been developed. Australia essentially escaped the Industrial Revolution that reshaped the societies of Europe, North America and Japan. Now the Australian government, in an extraordinarily enlightened policy decision, has decided to leapfrog over the industrial phase and embrace a new economic order based on the marketing of ideas, information and education. Prime Minister Bob Hawke has proclaimed that Australia can no longer be content to be "the lucky country," it has to become "the clever country."

The most tangible result, so far, of this decision is the plan to build a new type of city, known as a Multi-Function Polis (MFP), to be located near Adelaide. The MFP will involve research institutes, scientifically designed environmental schemes and social organizations, and advanced health, leisure and recreational facilities. There will be a strong emphasis on the networking concept, so that the MFP will consist of a collection of "villages" linked with high-tech optical communications. The MFP will in turn be networked to other cities, and ultimately to the rest of the world. The economic plan places a strong emphasis on ultrarapid communications and information networking, so that information, ideas and strategies can be marketed anywhere in the world, thus overcoming Australia's geographical isolation.

Perhaps the most imaginative component in the MFP strategy is the recognition that education and scientific research are highly valuable resources that can be marketed like any other. With a global network of communications, it would be possible for lectures given in Australia to be seen by students in the Third World, or for demonstrations of medical operations to be performed on one side of the planet and monitored by doctors on the other side. To achieve this goal the MFP will develop a "world university," by linking with local and more distant universities and educational institutions—the logical global development of the "Open University" pioneered in Britain two decades ago, using the basic communications systems then available.

These futuristic plans for Australia will surely become the norm throughout the world, as commodities assume less and less importance and ideas and information take their place. And the new social order will place its emphasis not on the clockwork image of Newtonian materialism, but on the network image of the post-Newtonian world view. For we live not in a cosmic clockwork, but in a cosmic network, a network of forces and fields, of nonlocal quantum connections and nonlinear, creative matter.

The nature of scientific truth

In the overthrow of the old world view—a paradigm shift that is dramatically transforming our understanding of reality—the chief casualty is common sense. Whereas in the Newtonian picture of reality human senses and intuition proved a good guide, in the abstract wonderland of the new physics it seems that only advanced mathematics can help us to make sense of nature. In breaking with Newtonian materialism we must accept that the objects of our theoretical models and the real entities of the external world bear a much more subtle relationship to each other than was assumed hitherto. Indeed, the very notion of what we mean by truth and reality must go into the melting pot.

In spite of our living in the so-called scientific age, science is not the only system of thought to command our attention. Various religions and alternative philosophies claim to offer a richer or more encompassing world view. The case for the scientific world view rests on the claim that science deals with *truth*. However elegant a scientific theory may be, and no matter how distinguished its originator, if it does not accord with experiment and observation it must be rejected.

This image of science as a pure and objective distillation of real world experience is, of course, an idealization. In practice, the nature of scientific truth is often much more subtle and contentious.

At the heart of the scientific method is the construction of *theories*. Scientific theories are essentially models of the real world (or parts thereof), and a lot of the vocabulary of science concerns the models rather than the reality. For example, scientists often use the word "discovery" to refer to some purely theoretical advance. Thus one often hears it said that Stephen Hawking "discovered" that black holes are not black, but emit heat radiation. This statement refers solely to a mathematical investigation. Nobody has yet seen a black hole, much less detected any heat radiation from one.

The relationship between a scientific model and the real system it purports to represent raises some deep issues. To illustrate the problem, we start with something fairly straightforward. In the sixteenth and seventeenth centuries the work of Copernicus, Kepler, Galileo and Newton overturned cen-

turies of entrenched ecclesiastical belief about the position of
the Earth within the Universe. Galileo was persecuted at the
hands of the church because he concurred with the Coperni-
can notion that the Earth moves around the Sun. This idea
conflicted with the then current theological interpretation of
biblical cosmology, which placed the Earth at the center of
creation.

It is a curious fact, however, that the church authorities did
not object to the concept of a moving Earth as such, so long
as it was only used as a model to compute the motions of
heavenly bodies. What they found intolerable was Galileo's
claim that the Earth *really* moves. But this raises an intriguing question. How is one to know when a scientific model is
merely a computational device and when it describes reality?

Science began as an extension of common sense, refined
and systematized to a high degree; so when scientists began
to build theories they usually started by taking the world at
face value. Thus, when ancient astronomers tracked the motion of the stars across the sky, they naturally devised a
model of the Universe in which the Earth was located at
the center of a collection of revolving spheres carrying the
Sun, Moon, stars and planets. As observations became
increasingly accurate, so this model had to be adapted
and readapted to include many spheres, and spheres
within spheres. This system of *epicycles* grew more and
more complicated. When Copernicus placed the Sun at the
center, the heavenly motions immediately became far simpler
to model.

Today, no scientist doubts that the Sun is *really* at the center of the Solar System, and that it is the Earth which re-
volves, not the sky. But is this certainty based merely on
Occam's razor—on the fact that the heliocentric model is
simpler than the geocentric model? Surely there must be
more to it than that?

Scientific theories are supposed to be *descriptions* of reality; they do not constitute that reality. It now seems obvious, though, that however successfully one fixed up the epicycle model to predict the positions of heavenly bodies, it would still be in some sense *wrong*. The problem is: how do we know that today's description of the Solar System is *right?* However certain we are that our present picture describes how the Universe *actually is,* we cannot rule out the possibility that some new and better way of looking at things, utterly unimaginable to us now, will be discovered in the future.

So long as scientific models stick closely to direct experience, where common sense remains a reliable guide, we feel confident that we can distinguish between the model and the reality. But in certain branches of physics it is not always so easy. The concept of energy, for example, is a familiar one today, yet it was originally introduced as a purely theoretical quantity in order to simplify the physicists' description of mechanical and thermodynamical processes. We cannot see or touch energy, yet we accept that it really exists because we are so used to discussing it.

The situation is even worse in the new physics, where the distinction between the model and reality sometimes becomes hopelessly blurred. In quantum field theory, for instance, theorists often refer to abstract entities called "virtual" particles. These ephemeral objects come into existence out of nothing, and almost immediately fade away again. Although a faint trace of their fleeting passage can appear in ordinary matter, the virtual particles themselves can *never* be directly observed. So to what extent can they be said really to exist? Might virtual particles be merely a convenient aid to the theorist's intuition—a simple way to describe processes that are otherwise unimaginable in terms of familiar concepts—

rather than real objects? Or might they be, like epicycles, an essential part of a model that will turn out to be wrong, and will be replaced by a model in which they have no place?

What is reality?

Generally, the more science moves away from common sense, the harder it is to decide what constitutes a mere model and what is supposed to be a faithful *description* of the real world. An outstanding mystery of particle physics is why the various subatomic particles have the masses they do. The proton, for example, weighs 1,836 times as much as the electron. Why 1,836? Nobody knows. A complete catalogue of all known particles would produce a list of several hundred such numbers. Although certain systematic trends can be discerned, the precise values of these numbers remain mysterious.

Now it is not inconceivable that somebody will one day invent a musical instrument that plays notes that have frequencies in precisely the same ratios as these peculiar numbers. The instrument would then be an excellent *model* for particle masses, but could one say that the particles *really are* notes in some abstract musical system? The idea seems ridiculous. But you have to be careful. As we have mentioned, physicists are currently very excited about the theory of superstrings, which claims that what we have always thought of as subatomic particles are actually excitations—or vibrations—of little loops of string! So the instrument idea may not be so crazy after all. On the other hand, we cannot actually observe these strings—they are far too small. So should we think of them as *real* or as only a theoretical construct?

If history is anything to go by, nature has a nasty habit of deceiving us about what is real and what is invented by human beings. The apparent motion of the stars, reflecting the real motion of the Earth, is only one of a long list of examples in which scientists have been led astray by taking nature too much at face value.

Many other examples come from biology. Biological organisms are so remarkable in their properties it is easy to suppose that they are infused with some special substance, or life force. This theory is known as vitalism, and it was very popular in the early part of the twentieth century. Hans Dreisch, for example, was greatly impressed by the way that embryos develop from a simple egg into the elaborately organized form of the advanced fetus. What he found particularly mysterious was the ability of some embryos to recover from deliberate mutilation. It seemed to Dreisch that this unfolding of order must be under the supervision of some unseen guiding force, which he called *entelechy*.

Today vitalism is totally discredited. Advances in molecular biology, such as the unraveling of DNA and the cracking of the genetic code, have demonstrated that life is based on chemical reactions that do not differ in any fundamental respect from those that take place in inanimate systems. Dreisch and others were misled, it now seems, by their understandable failure to appreciate how large numbers of molecules can act together in an apparently cooperative way without, in fact, being coordinated by any imposed plan.

The history of the theory of evolution is littered with similar pitfalls. Consider, for example, how plausible Lamarck's theory of the inheritance of acquired characteristics seems. Organisms are continually striving toward goals: lions are trying to run faster, in order to catch their prey; giraffes are straining their necks to reach higher foliage; and so on. These experiences, claimed Lamarck, have an effect on their offspring, so that the next generation of lions can run a little faster and the next generation of giraffes are born with slightly longer necks. The son of a blacksmith, according to Lamarck, will be born with a tendency to develop large muscles, because his father has used the equivalent muscles in his working life. In this way, species become adapted more and more successfully to their environment.

Lamarck's theory has a considerable appeal to common sense. One has only to observe living things to be persuaded that they are striving, and we know from the fossil record that species become progressively better adapted to their individual ecological niches as the generations pass. Nevertheless, the theory is wrong. Experiment and observation show that the characteristics acquired by an organism during its lifetime (such as a blacksmith's big muscles) are not passed on genetically to its offspring. Instead, as Darwin correctly surmised, changes take place from one generation to the next entirely at random, and it is natural selection that preserves the advantageous mutations and thereby brings about the progressive nature of evolutionary change.

The philosopher Thomas Kuhn believes that scientists adopt certain distinct paradigms that are tenaciously retained and are abandoned only in the face of glaring absurdities. These paradigms help to shape scientific theories, and exercise a powerful influence over the methodology of science and the conclusions drawn from experiments. Experimental scientists pride themselves on their objectivity, yet time and again they unwittingly massage their data to fit in with preconceived ideas. Sometimes, several different independent experimenters will carefully measure the same quantity and consistently get the same *wrong* answer, because it is the answer they have come to expect.

The canals of Mars are a case in point. After G. V. Schiaparelli had reported, in 1877, seeing a network of lines on the

Martian surface, many other astronomers confirmed their existence, even to the extent of producing detailed maps. When the Mariner 4 spacecraft sent back the first detailed photographs of Mars during its 1965 fly-by, however, there was no trace of the "canals."

Or take the phlogiston theory of combustion. In the seventeenth century Georg Ernst Stahl proposed that when a material burns or rusts it is giving off a substance called phlogiston. The idea seemed natural—a burning or rusting object does look as if it is giving up something to the air. But once again, appearances proved misleading. Later studies showed that combustion and rusting involve taking something (oxygen) *from* the air.

This is a nice example of scientists seeing things that aren't there. In other cases they fail to see things that *are* there. The existence of meteorites was doubted for many years—it was regarded as scientific nonsense to suggest that rocks could fall from the sky. Then, one particularly spectacular fall in France forced the French Academy to change its position, and the rest of the scientific community soon followed.

Beyond common sense

When a paradigm shift occurs in science, it is often accompanied by enormous controversy. A classic example concerns the "luminiferous ether." When James Clerk Maxwell showed that light is an electromagnetic wave, it seemed obvious that this wave had to have a medium of some sort through which to propagate. After all, other known waves travel *through* something. Sound waves, for example, travel through the air; water waves travel across the surface of lakes and oceans. Because light, which Maxwell discovered is a form of electromagnetic wave, can reach us from the Sun and stars, across seemingly empty space, it was proposed that space is actually filled with an intangible substance, the ether, in which these waves could travel.

So sure were physicists of the existence of the ether that ambitious experiments were mounted to measure the speed with which the Earth moves through it. Alas, the experiments showed conclusively that the ether does not exist. This roused fierce debate, until the dilemma was resolved, in 1905, by a paradigm shift. By supposing that space and time are elastic, and change from one reference frame to another, Einstein was able to demonstrate that his theory of relativity rendered the ether superfluous. Instead, light was treated as a wavelike disturbance in an independently existing electromagnetic field. The field transforms from one reference frame to another in such a way that the Earth's motion is irrelevant.

For nineteenth-century physicists, however, the ether was still very real. Indeed, some people (not physicists, though!) cling to the idea today. One still hears talk about radio transmissions as "waves in the ether." But this is largely just a figure of speech. The question is, how can we be sure there isn't an ether? After all, the electromagnetic field is also an abstract entity that we cannot directly observe. One can point again to the fact that the relativistic field theory is simpler than the alternative. But whereas the issue seems clearcut in the case of the Earth going around the Sun, the question of whether the ether, or the electromagnetic field, or neither, is "really there" seems altogether more subtle.

So determined are some people to hang on to a "commonsense" view of reality that they challenge even the most firmly established ideas of the new physics. Einstein's theory of relativity, with its counterintuitive notions of space and time, attracts particular attention. Even after nearly a century of careful testing of this theory, editors of scientific journals continue to be deluged with papers (most of them from authors with minimal scientific training) purporting to find flaws in Einstein's work and attempting to return us to the safe old world of absolute space and time. The usual motivation for these misguided attacks is that the world cannot "really" be as Einstein claimed, that any theory dealing with "the truth" must be comprehensible in simple terms and not require abstract models.

The difficulties concerning the relationship between abstract models and reality do not, however, undermine the claim that science deals with truth. Clearly, scientific theories—even in the most abstract form—capture some element of reality. But one may certainly question whether science can deliver *the whole truth*. Of course, many scientists deny that science ever makes such a grandiose claim. Science may be very good at explaining, say, electrons, but it has limited utility when it comes to things like love, or morality, or the meaning of life. These experiences are still part of our reality, but they seem to lie beyond the scope of science.

It may be that the failure of science to deliver on these deep issues of existence has led to the widespread disillusionment with the scientific world view that is fueling the current antiscience backlash in Western society. The danger is that science will be rejected in favor of other systems of thought based more on dogma than empiricism. Worse still is the growing tendency for science to be retained as a procedure, but distorted or manipulated to fit in with certain preconceived doctrines. Witness, for example, the rise of so-called creation science, and more recently "Islamic science" and "feminist science." There is, of course, only *science,* and it deals with truth, not dogma. The important thing is to appreciate that this truth may be limited and fail to satisfy the desire of some people to grasp the ultimate reality.

It may be wondered whether science will always be limited in this respect. Is it possible to imagine that future developments will enable science to answer the ultimate questions and to deal with the total reality? The answer would seem to be no, for remarkably science contains within itself a description of its own limitations.

In the 1930s physicists were strongly influenced by a philosophical movement known as positivism, which seeks to root reality only in what can actually be observed. The founders of quantum mechanics, notably Niels Bohr and Werner Heisenberg, argued that when we talk of atoms, electrons, and so on, we must not fall into the trap of imagining them as little "things," existing independently in their own right. Quantum mechanics enables us to relate different *observations* made on, say, an atom. The theory is to be regarded as a procedure for connecting these observations into some sort of consistent logical scheme—a mathematical algorithm. Use of the word "atom" is just an informal way of talking about that algorithm. It is a helpful means of encapsulating that abstract concept in physical language, but that does not mean that the atom is actually *there* as a well-defined entity with a complete set of physical attributes of its own, such as a definite location in space and a definite velocity through space.

Heisenberg's own words are revealing in this context: "In the experiments about atomic events we have to do with things and facts, with phenomena that are just as real as any phenomena in daily life. But the atoms or the elementary particles themselves are not as real; they form a world of potentialities or possibilities rather than one of things or facts." Bohr expressed it thus: "Physics is not about how the world *is*, it is about what we can *say* about the world." For these physicists, reality did not go beyond the facts of experience, the results of measurements made on pieces of macroscopic apparatus. The term "atom" itself became merely a code word for a mathematical model. It was not intended to represent an independent part of reality.

Not all physicists have been prepared to accept this position. Einstein, for example, opposed it implacably. He insisted that the quantum microworld contains objects such as atoms that are every bit as real as tables and chairs. They differ, he argued, only in scale from the objects of everyday experience in this respect. This dissenting tradition has been kept alive by David Bohm, who continues to argue that there *is* a reality in the microworld, even though our observations at present reveal it only imperfectly.

These deep divisions within the scientific community, concerning the nature of reality, point up the shakiness of any claim that science deals with the whole truth. Quantum mechanics seems to impose an inherent limitation on what science can tell us about the world, and it reduces to mere models entities that we used to regard as real in their own right.

In spite of the widespread support given to the philosophy of Bohr and Heisenberg, the desire to ask what is *really* the case in the world is overwhelming. Do atoms really exist? Does the ether really exist? The answers seem to be, respectively, "perhaps," and "probably not"; but science can never tell us.

Faced with this limitation, some people may prefer to reject science altogether and rely on religion, or to embrace one of the wilder modern schemes, such as scientology, creationism or the ideas of von Daniken. But this would be a grave mistake. It is surely better to accept a system of thought that sets uncompromising standards of skepticism and objectivity, even if it can only provide a partial description of reality, than to retreat into an uncritical acceptance of dogma. That is not to say that there is no place for religion, of course—so long as religion restricts itself to those questions that lie outside the scope of science. Indeed, for many people those will be the questions that really matter.

But this is enough about the limitations of science. Having tried to be honest about what science cannot tell us about the Universe, from now on we intend to describe what science can tell us about the world we inhabit, and the new reality that is emerging from the modern understanding of the behavior, not of individual "atoms" and "particles" (however real or unreal they may be), but of aggregates of many particles operating—or cooperating—in complex systems. The paradigm shift that we are now living through is a shift away from reductionism and toward holism; it is as profound as any paradigm shift in the history of science.

2 Chaos and the Liberation of Matter

All science is founded on the assumption that the physical world is ordered. The most powerful expression of this order is found in the laws of physics. Nobody knows where these laws come from, nor why they apparently operate universally and unfailingly; but we see them at work all around us, in the rhythm of night and day, the pattern of planetary motions, or the regular ticking of a clock.

The ordered dependability of nature is not, however, ubiquitous. The vagaries of the weather, the devastation of an earthquake and the fall of a meteorite seem to be arbitrary and fortuitous. Small wonder that our ancestors attributed these events to the whim of the gods. But how are we to reconcile these seemingly random "acts of God" with the supposed underlying lawfulness of the Universe?

The Ancient Greek philosophers regarded the world as a battleground between the forces of order, producing cosmos, and those of disorder, which worked toward chaos. Random or disordering processes were seen as negative, evil influences. Today, we do not regard the role of chance in nature as malicious, merely as blind. It may act constructively, as in biological evolution, as well as destructively, as when an aircraft wing fails from metal fatigue.

Though individual chance events may give the impression of lawlessness, disorderly processes may still display deep statistical regularities. Indeed, casino managers put as much faith in the laws of chance as engineers put in the laws of physics. But this raises something of a paradox. How can the same physical processes—such as the spin of a roulette wheel—obey both the laws of physics and the laws of chance?

Is the Universe really a machine?

As we have seen, following the formulation of the laws of mechanics by Isaac Newton in the seventeenth century, scientists became accustomed to thinking of the Universe as a gigantic mechanism. The most extreme form of this doctrine was strikingly expounded by Pierre Laplace in the nineteenth century. He envisaged every particle of matter as unswervingly locked in the embrace of strict mathematical laws of motion. These laws dictated the behavior of even the smallest atom in the most minute detail. Laplace realized that if this were so, then, given the state of the Universe at any one instant, the entire cosmic future would be uniquely fixed, to infinite precision, by Newton's laws.

The concept of the Universe as a strictly deterministic machine governed by immutable laws profoundly influenced the scientific world view, as we mentioned in Chapter 1. It stood in stark contrast to the old Aristotelian picture of the cosmos as a living organism. A machine can have no "free will"; its future is rigidly determined from the beginning of time. Indeed, time ceases to have much physical significance in this picture, for the future is already contained in the present (and so, for that matter, is the past). As Ilya Prigogine has eloquently expressed it, God is reduced to a mere archivist, turning the pages of a cosmic history book that is already written.

Implicit in this somewhat bleak mechanistic picture was the belief that there are actually no truly chance processes in nature. Events may seem to us to be random, but, it was reasoned, this can always be attributed to human ignorance about the details of the processes concerned.

Take, for example, Brownian motion. A tiny particle suspended in a fluid (even a dust mote floating in the air) can be seen, using a microscope, to execute a haphazard zigzag movement as a result of the slightly uneven buffeting it suffers at the hands of the molecules of the fluid that bombard it on all sides. Brownian motion is the archetypal random, unpredictable process. Yet, so the Laplacian argument ran, if we were able to follow in detail all the activities of all the individual molecules involved, Brownian motion would be every bit as predictable and deterministic as clockwork. The seemingly random motion of the Brownian particle is attributed solely to the lack of information about the myriads of participating molecules, a lack attributable to the fact that our senses (and our measuring instruments) are too coarse to permit detailed observation at the molecular level.

For a while, it was commonly believed that apparently "chance" events were always the result of our ignoring, or effectively averaging over, vast numbers of events occurring at this hidden level. The toss of a coin or the roll of a die, the spin of a roulette wheel—these would no longer appear random, it was thought, if we could observe the molecular world. The slavish conformity of the cosmic machine would ensure that lawfulness was folded up in even the most haphazard events, albeit in an awesomely convoluted tangle.

Two major developments of the twentieth century have, however, laid to rest the idea of a clockwork Universe. First there is quantum mechanics. At the heart of quantum mechanics lies Heisenberg's uncertainty principle, which states that everything we can measure is subject to truly random fluctuations. More of this in Chapter 7; the essential point is that quantum fluctuations are not the result of human limitations or hidden levels of mechanistic clockwork; they are *inherent* in the workings of nature on an atomic scale. For example, the exact moment of decay of a particular radioactive nucleus is intrinsically uncertain. An element of genuine unpredictability is thus an integral part of nature.

In spite of this uncertainty, there remains a sense in which quantum mechanics is still a deterministic theory. Although the outcome of a particular quantum process might be undetermined, the *relative probabilities* of different outcomes evolve in a deterministic manner. What this means is that although you cannot know in any particular case what will be the outcome of the "throw of the quantum dice," you can know completely accurately how the betting odds vary from moment to moment. As a *statistical* theory, quantum mechanics remains deterministic. This is why a machine such as a computer or a CD player, which depends for its functioning on the behavior of innumerable quantum particles, such as electrons, following the statistical rules, can still function reliably even though the behavior of each individual electron within the machine cannot be predicted. Quantum physics builds chance into the very fabric of reality, but a vestige of the Newtonian-Laplacian world view remains.

Then along came chaos. The essential ideas of chaos were already present in the work of the French mathematician Henri Poincaré at the end of the nineteenth century; but it is only in recent years, especially with the advent of fast electronic computers with which to carry out the appropriate calculations, that the full significance of chaos theory has been appreciated.

The key feature of a chaotic process concerns the way that *predictive errors* evolve with time. To explain this, we can start with an example of a nonchaotic system: the motion of a simple pendulum. Imagine two identical pendulums swinging exactly in step with each other—in synchronism. Suppose that one pendulum is slightly disturbed, so that its motion gets a little out of step with the other pendulum. This discrepancy (termed a phase shift) remains small as the pendulums go on swinging.

Faced with the task of predicting the motion of a simple pendulum, one could measure the position and velocity of the bob at some instant and use Newton's laws to calculate the subsequent behavior. Any error in the initial measurement propagates through the calculation and appears as an error in the prediction. For the simple pendulum, a small input error implies a small output error in the predictive computation. The phase shift between the two swinging pendulums gives a picture of such an "error" at work.

In a typical nonchaotic system, errors accumulate with time. Crucially, though, the errors grow roughly in proportion to the time that has elapsed since the prediction was made, so they remain relatively manageable.

Now contrast this behavior with that of a chaotic system. In such a system, any small starting difference between two identical systems will grow rapidly. In fact, the hallmark of chaos is that the two motions diverge *exponentially* fast. In the language of our predictive problem, this means that any input error increases at an escalating rate. Instead of the error growing by roughly the same amount with each second that passes, it may grow by as much in each successive second as in all the previous seconds together since the predictive sequence was started. Before long, the error engulfs the calculation, and all predictive power is lost. Small input errors thus swell to calculation-wrecking size in very short order.

The distinction between these two modes of behavior is well illustrated by the behavior of a spherical pendulum— one that is free to swing in any direction. In practice, this could be a ball suspended on a piece of string from a pivot (Figure 1). If the system is driven with a smooth, periodic,

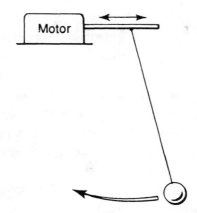

Figure 1. Even a simple "spherical" pendulum can exhibit chaos. When the top end of the string is oscillated, the ball swings about. For some frequencies the ball settles into a regular orbital motion. For other oscillation frequencies the ball's motion is so erratic that it is essentially random and unpredictable.

horizontal motion at the pivot, the ball will start to swing about. After a while, it may settle down into a stable and entirely predictable pattern of motion in which the bob traces out a roughly elliptical path, with the same period as the driving force. But if the driving frequency is altered slightly, this regular motion may (if the frequency is close to some critical value) give way to chaos, with the bob swinging first this way and then that, doing a few clockwise turns, then a few anticlockwise turns, and so on, in a random manner.

The randomness of this system does not arise, as is the case with Brownian motion, from the effect of myriads of interactions operating at a hidden level—what physicists call "hidden degrees of freedom." Indeed, we can describe the system using a mathematical model in which there are only three degrees of freedom, and the model itself is strictly deterministic. Nevertheless the behavior of the pendulum described by the model is random. It used to be supposed that determinism goes hand in hand with predictability. The chaotic pendulum demonstrates that this is not necessarily the case.

A deterministic system is one in which future states are completely determined, through some dynamical law, by preceding states. Drop a ball, and it falls to the ground. The position and velocity of the ball at any moment during its fall are completely fixed by its position and motion at the moment of release. There is thus a one-to-one association between earlier and later states. In computational terms, this suggests a one-to-one association between the input and the output of a predictive calculation. But now we must remember that any predictive computation will necessarily contain *some* input errors, because we cannot measure physical quantities to unlimited precision. Moreover, a computer is capable of handling only finite quantities of data.

The distinction between nonchaotic and chaotic systems may then be illustrated schematically by analogy with two different geometrical constructions. In Figure 2, points on the top horizontal line represent the starting conditions of a nonchaotic system (for example, the position of a ball about to be dropped). Points on the bottom horizontal line represent the state of the system at some later time (such as the position of the ball one second after it has started its fall). Determinism means that there is a one-to-one correspondence between points on the top line and points on the bottom line, a correspondence represented here by vertical lines. Each final state (each point on the bottom line) is reached from one and only one initial state (a point on the top line). If we are slightly ignorant about the initial state, this ignorance translates into slight ignorance about the final state. On the diagram, this corresponds to the fact that

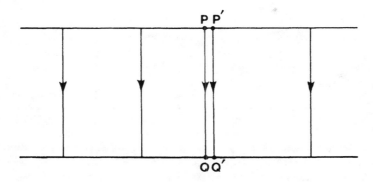

Figure 2. Determinism can be demonstrated symbolically using this simple geo-metrical construction. Each point on the top line is uniquely associated with a point on the bottom line, via a vertical line. Point P, for example, is associated with point Q. A point P′ close to P will be associated with a point Q′ close to Q, and so on. Small errors in our knowledge about the position of P correspond to only small errors in our knowledge about the position of Q. If points on the upper and lower lines represent initial and final states of a physical system, then this construction symbolizes predictability.

closely spaced points on the top line are connected to closely spaced points on the bottom line. Thus a small error in our knowledge of the initial state implies only a small error in the predicted final state.

In the case of a chaotic system the situation resembles that shown in Figure 3. Here, the initial states are represented by points on the arc of a circle, while final states are represented by points on the horizontal line. Again there is a one-to-one correspondence between these two sets of points: given a point on the arc, a point on the straight line is uniquely determined. But in this case the lines that connect these two sets of points fan out, so that as the top of the arc is approached the corresponding points on the straight line become progressively more spaced out. Very slight changes in the starting point yield dramatically different end points,

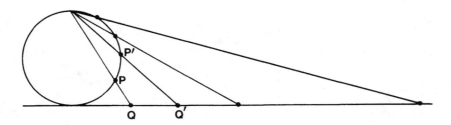

Figure 3. In contrast to the situation depicted in Figure 2, in this representation nearby points P and P′ on the arc are associated with more widely separated points Q and Q′ on the horizontal line. The rule for locating Q is that it lies on the horizontal line at the point where a line drawn from the top of the arc through P crosses the horizontal line. The sensitivity becomes more pronounced as P gets closer to the top of the arc. In spite of the fact that points on the horizontal line are uniquely determined by those on the arc, slight errors in the location of P produce big errors in the location of Q. The relationship is deterministic, but difficult to predict.

so that slight ignorance about the initial state now leads to great uncertainty concerning the final state. This situation symbolizes chaos, where the system is incredibly sensitive to the initial conditions, and very slightly different starting states lead to dramatically different end states.

This sensitivity is not just a result of human inability to calculate with enough precision or to draw fine enough lines. The mathematical concept of a line is a kind of fiction, an approximation to reality. It is the uncertainty that is real, and the idealized mathematical line that is the fiction. We can see this clearly by looking at a mathematical description of lines developed by the Ancient Greeks.

They realized that one could label points on a line by numbers, indicating the distance of each point from the end of the line. Figure 4 shows a segment from 0 to 1. To label the points in between, we can use fractions, such as ⅔ or ¹³⁷/₅₅₄. The Greeks called these numbers "rational," from the

Figure 4. Points on a line can be used to represent numbers between 0 and 1. There are an infinite number of fractions in this interval, but even so there are not enough fractions to label every point on the line.

root "ratio." By using enough digits in the numerators and denominators we can choose a fraction that marks a place arbitrarily close to any designated point on the line. Nevertheless, mathematicians can easily prove that a continuous line segment cannot have all its points labeled by rational numbers. You can get as close as you like to one of these "extra," or "irrational," points by choosing a suitable such fraction, but you can never find a fraction that lands you precisely on an irrational point. To label every point on the line, you need not only all possible rational numbers, but all irrational numbers as well. An irrational number cannot be expressed as one whole number divided by another whole number, but it may instead be expressed as a decimal, with an infinite number of digits after the decimal point.

The set of all rational and irrational numbers forms what mathematicians call the real numbers, and they underlie almost all modern physical theory. The very notion of continuous mechanical processes, epitomized by Newton's calculus (which he formulated to describe such processes), including the fall of an apple from a tree or the orbit of the Moon around the Earth, is rooted in the concept of real numbers. Some real numbers can be expressed compactly. These include ½ = 0.5 and ⅓ = 0.333 . . . But a typical real number can only be expressed as a decimal expansion consisting of an infinite string of digits with no systematic pattern to it. It is a random sequence. It follows that to specify even *one* such number involves an *infinite* quantity of information. This is clearly impossible, even in principle. Even if we were to commandeer the entire observable Universe and use it as a digital computer, its information storage capacity would still be finite, and it could not even "remember" *one* irrational number with complete precision. Thus, the notion of a continuous line described by real numbers is exposed as a mathematical fiction.

Consider the consequences of this for a chaotic system. Determinism implies predictability only in the idealized limit of infinite precision. In the case of the pendulum, for example, the behavior will be determined uniquely by the initial conditions. The initial data include the position of the bob, so exact predictability demands that we must assign the real number that correctly describes the distance of the center of the bob from a fixed reference point. This infinite precision is, as we have seen, impossible.

In a nonchaotic system this limitation is not so serious because the errors grow only slowly. But in a chaotic system errors grow at an accelerating rate. Suppose there is an uncertainty in, say, the fifth significant figure of the relevant decimal expression, and that this affects the prediction of how the system is behaving after a time t. A more accurate analysis might reduce the uncertainty to the tenth significant figure. But the exponential nature of the error growth implies that the uncertainty is already back to its old value after a time $2t$. Increasing the initial accuracy by a factor of 100,-000 only doubles the predictability span. And the situation is the same whether we are talking about mathematical calculations or about some small physical disturbance of the system from outside, like the disturbance that creates a phase difference in the swinging of two identical pendulums, or sets a spherical pendulum off into chaotic motion.

It is this "sensitivity to initial conditions" that leads to the well-known statement that the flap of a butterfly's wings in Adelaide today can affect the weather in Sussex next week. Because the Earth's atmosphere is a chaotic system, and because no system can *in principle* be described with perfect precision, completely accurate long-term weather forecasting can never be achieved—nor can accurate forecasting of any other chaotic system. We stress that this is not just a human limitation. The Universe itself cannot "know" its own workings with absolute precision, and therefore cannot "predict" what will happen next, in every detail. Some things really are random.

Chaos evidently provides us with a bridge between the laws of physics and the laws of chance. In a sense, chance, or random, events can indeed always be traced to ignorance about details. But whereas Brownian motion seems random because of the enormous number of degrees of freedom we are voluntarily overlooking, deterministic chaos seems random because we are *necessarily* ignorant of the ultrafine detail of just a few degrees of freedom, *and so is the Universe itself.* And whereas Brownian motion is complicated because the molecular bombardment of the dust mote is itself a complicated process, the motion of, say, a spherical pendulum is complicated even though the system itself is very simple. Thus, complicated behavior does not necessarily imply complicated forces or laws. The study of chaos has revealed how it is possible to reconcile the complexity of a physical world displaying haphazard and capricious behavior with the order and simplicity of the underlying laws of nature.

Though the existence of deterministic chaos comes as a surprise, it should not be forgotten that nature is not, in fact, deterministic anyway. The indeterminism associated with quantum effects will intrude into the dynamics of all systems, chaotic or otherwise, at an atomic level. It might be supposed that quantum uncertainty would combine with chaos to amplify the unpredictability of the Universe. Curiously, though, quantum effects seem to have a subduing effect on chaos. Some model systems that are chaotic at the "classical" level of Newtonian mechanics are found to be nonchaotic when quantized. At this stage, experts are divided about whether quantum chaos is possible, or what its signature might be if it does exist. Though the topic will undoubtedly prove important for atomic and molecular physics, it is of little relevance to the behavior of macroscopic objects, let alone the Universe as a whole.

What can we conclude about the Newtonian-Laplacian image of a clockwork Universe? The physical world contains a wide range of both chaotic and nonchaotic systems. Weather, as we have mentioned, is inherently unpredictable in its fine detail, but the march of the seasons really does seem to be as regular as clockwork. Those systems that are chaotic have severely limited predictability, and even one such system would rapidly exhaust the capacity of the entire Universe to compute its behavior. It seems, then, that the Universe is incapable of computing the future behavior of even a small part of itself, let alone all of itself. Expressed more dramatically, the Universe is its own fastest simulator.

This is surely a profound conclusion. It means that, even accepting a strictly deterministic account of nature, the future states of the Universe are in some sense "open." Some people have seized on this openness to argue for the reality of human free will. Others claim that it bestows upon nature an element of creativity, an ability to bring forth that which is genuinely new, something not already implicit in earlier states of the Universe. Whatever the merits of such claims, it seems safe to conclude from the study of chaos that the future of the Universe is not irredeemably fixed. To paraphrase Prigogine, the final chapter of the great cosmic book has yet to be written.

Comprehending complexity

The astonishing success of simple physical principles and mathematical rules in explaining large parts of nature is not something that is obvious from our everyday experience; nor was it obvious to our ancestors that the world runs on such simple lines. On casual inspection nature seems dauntingly complex and utterly beyond comprehension. Few natural phenomena overtly display any very precise sort of regularity that might hint at the underlying order. Where trends and rhythms are apparent, they are usually of an approximate, qualitative form. Indeed, centuries of careful investigation by Ancient Greek and Medieval thinkers failed to uncover any but the most trivial examples (such as the cycle of day and night) of an underlying mathematical order in nature.

The situation can be highlighted by looking at the example of falling objects. Galileo realized that all bodies accelerate at the same rate in the Earth's gravity. Nobody had realized this before, because in everyday experience it simply is not true. Everybody knows that a hammer falls faster than a feather. Galileo's genius lay in spotting that the differences that occur in the everyday world are an incidental complication (in this case, caused by air resistance) and are irrelevant to the underlying properties (that is, gravity). He was thus able to abstract from the complexity of real-life situations the simplicity of an idealized law of gravity.

The work of Galileo and Newton in the seventeenth century is often taken to mark the beginning of modern science. The success of science is largely based on the power of the kind of analysis used by Galileo—the practice of isolating a physical system from the outside Universe and focusing attention on the phenomenon of interest. In the case of falling objects, such isolation might involve experimenting in a vacuum, for example; nobody who saw it can fail to have been impressed when the Apollo astronauts actually carried out Galileo's hypothetical experiment, dropping a feather and a hammer together on the airless Moon and watching them fall at the same rate.

But the fact that this kind of analysis works is itself something of a mystery. The world, after all, is an interconnected whole. Why is it possible to know something, such as the law of falling objects, without knowing everything? Indeed, why is it possible to know *so much* without knowing everything?

If the Universe were an "all-or-nothing" affair, there would be no science and no understanding. We could never apprehend all the principles of nature in a single grasp. And yet, in spite of the widespread belief these days among physicists that all the principles will indeed turn out to form a coherent unity, we are nevertheless able to proceed one step at a time, filling in small areas of the jigsaw puzzle without needing to know in advance the finished picture that will appear. This has happened throughout the three and a half centuries of scientific endeavor. At a more personal level, it happens to each new, would-be scientist setting out on the necessary fifteen years or so of education. To be a scientist, you do *not* have to comprehend all of modern physics in one swallow!

Part of the reason for the success of the step-by-step approach is that many physical systems are approximately linear in nature. In physics, a linear system is, simply speaking, one in which the whole is equal to the sum of its parts (no more, no less), and in which the sum of a collection of causes produces a corresponding sum of effects.

The distinction between linear and nonlinear relationships can be simply illustrated by the example of a sponge soaking up water. If water is dripped onto a dry sponge, the weight of the sponge will increase. The weight increase will at first be proportional to the number of drips: twice the number of drips causes twice the increase in weight. This is a linear relationshp. But when the sponge gets very wet, it will start to saturate. In this state, its capacity to absorb more water will be reduced, and some of the water that drips onto it will run off again. As its capacity to absorb water declines, the weight increase becomes nonlinear—progressively *less,* in this example, for each additional drip. Eventually the weight will stabilize and become independent of the number of additional drips, because each new drip being added to the sponge is balanced by the same amount of escaping water. This behavior is illustrated in Figure 5.

A complicated linear system, such as a radio wave that is modulated by the sound of a voice, can be separated into components (in this case, different waveforms) and put back together again without introducing any distortion—the complex waveform simply consists of a lot of different simple waveforms superimposed. The very concept of scientific "analysis" depends on this property of linearity—that understanding the parts of a complex system implies understanding the whole. And this ability to "decompose without destroying" a linear system is reflected in the mathematics that is used to describe the system. Linear mathematics is especially tractable because its complexity can likewise be analyzed into the superposition of simple expressions.

The success of linear methods over the past three centuries has, however, tended to obscure the fact that real systems almost always turn out to be *nonlinear* at some level. When nonlinearity becomes important, it is no longer possible to

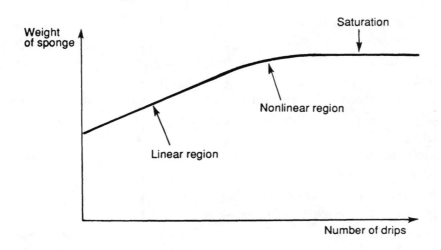

Figure 5. For a sponge that starts out dry and has water dripped steadily onto it, the relationship between the weight of the sponge and the number of drips is called linear because the graph relating the two is a straight line. When the sponge starts to saturate, the relationship becomes nonlinear.

proceed by analysis, because the whole is now greater than the sum of its parts. Nonlinear systems can display a rich and complex repertoire of behavior, and do unexpected things—they can, for example, go chaotic. Without nonlinearity, there would be no chaos, because there would be no diversity of possible patterns of behavior on which the intrinsic uncertainty of nature could act.

Generally speaking, a nonlinear system must be understood in its totality—which in practice means taking into account a variety of constraints, boundary conditions and initial conditions. These supplementary aspects of the problem must be included in the study of linear systems, too; but there they enter in a rather trivial and incidental way. In the case of nonlinear systems, they are absolutely fundamental in determining what is going on.

We have already seen an example of this in the previous section. The determining factor of whether or not a pendulum goes chaotic concerns the frequency of the external driving force in relation to the length of the pendulum. The whole system has to be taken into account before we can predict the onset of chaos. There are many other examples of what might be called the holistic character of nonlinear systems. These include self-organizing phenomena, such as chemical mixtures that grow shapes or pulsate with patterns of color in cooperative ways. A full discussion of the significance of these ideas can be found in *The Cosmic Blueprint;* here we particularly wish to point out that an understanding of the *local* physics (such as the forces between molecules) may be *necessary* to understanding what is going on, but it is certainly not *sufficient* to explain the phenomena fully.

The nonlinearity of physical systems bestows upon them an uncanny ability to do unexpected things, sometimes with an almost lifelike quality. They may behave cooperatively, spontaneously adapt to their environment or simply arrange themselves into coherent entities with a distinct identity. We are a world away from Newton's inert matter, as we can illustrate by taking a close look at just one example of this "liberation" of matter from its lumpen mold. This is one of the most important examples of nonlinearity at work: nonlinear waves.

Waves with a will of their own
In the year 1834 an engineer by the name of John Scott Russell was out riding near Edinburgh when he came upon a boat being pulled along a narrow canal by a pair of horses. As Russell watched, the boat came to an abrupt stop, creating a violent disturbance in the water. To Russell's astonishment, a large hump of water rose up at the bow of the boat

and, as he later wrote, "rolled forward with great velocity, assuming the form of a large solitary elevation, a rounded, smooth and well-defined heap of water, which continued on its course along the channel apparently without change of form or diminution of speed." Russell, mounted on his horse, chased after the enigmatic phenomenon for two miles before losing it in the windings of the channel.

We are all familiar with ordinary water waves, but the spectacle witnessed by Russell was something quite out of the ordinary. If a stone is dropped in a pond, the ripples that are created spread out from the point of impact and gradually die away. Unlike these ordinary water waves, which form a succession of peaks and troughs, Russell's "heap of water" was a single hump in the surface that retained its identity as it propagated. Such "solitary waves" are by no means a freak occurrence; indeed, Russell returned many times to the canal to study the phenomenon, and wrote a report on his research for the *Transactions of the Royal Society of Edinburgh*.

It was not until 1895, however, that two Dutch physicists, D. J. Korteweg and Hendrik de Vries, were able to provide a satisfactory explanation for solitary waves of the sort witnessed by Russell. Their theory—or close variants of it—now finds applications in many other areas of science, from particle physics to biology.

To understand the theory, it is first necessary to know something about ordinary waves. A typical group of water waves, such as the ripples that result from throwing a stone in a pond, consists of a train of undulations. A wave group like this is actually made up of a lot of waves of different amplitude (different heights) and lengths (the wavelength is the distance from one peak to the next) superimposed on one another. Near the center of the group all these contributing waves are more or less in step (in phase) and so they reinforce each other to make a large disturbance. Near the edges of the wave group, the various contributing waves get out of step, because of their different lengths, and they tend to cancel each other, thus reducing the disturbance. The effect is to bunch the undulations into a confined region.

The particular shape of the wave group depends on the specific mixture of waves present. As the wave group progresses, however, its state changes. This is because long waves on water travel faster than short waves. The phenomenon is called *dispersion,* because it causes the wave group to spread out and eventually fade away.[1]

For a single hump of water to propagate in such a dispersive medium without spreading out and fading away requires another factor to be at work to counteract the effects of dispersion. This new factor is an example of nonlinearity at work. The sort of waves we usually witness are called linear waves because of the way they combine. Any two linear waveforms, when superimposed, produce a merged wave in which the amplitude at each point is obtained from the amplitudes of the two original waves simply by adding them together (Figure 6). For this to apply, it is necessary that the speed of each wave, although depending on its length, does not depend upon its *height*. If it does, then adding two waves together changes the speed with which the combined wave propagates, and the picture becomes more complicated. Korteweg and de Vries realized that the assumption of linearity is valid for water waves only if their amplitude is small compared with the depth of water. If the water is shal-

---

1 The same effect involving light waves traveling through lenses produces colored fringes to images viewed through a telescope, a bane of early astronomical observations.

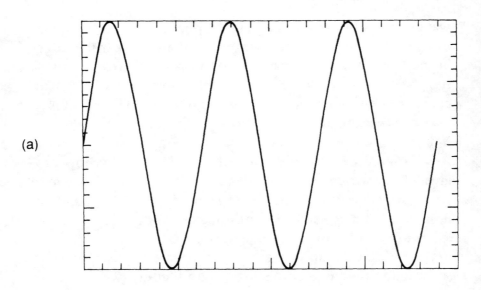

(a)

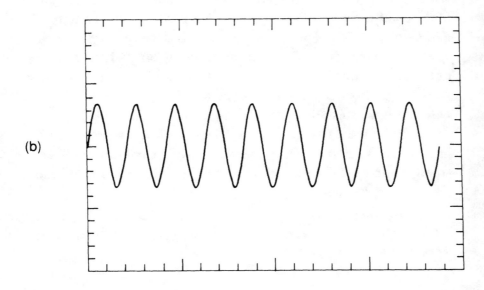

(b)

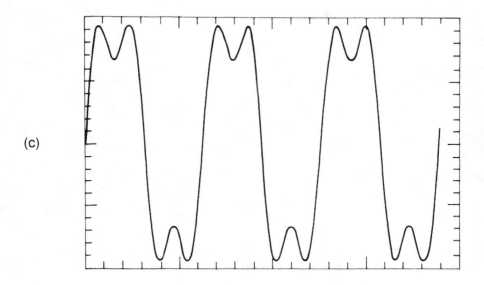

(c)

Figure 6. Linear waves may be superimposed just by adding their amplitudes together at each point. Thus wave *a* and wave *b* combine to give wave *c*. Nonlinear waves combine in more complicated ways.

low and the wave amplitude rather large in comparison, the speed of the waves depends *both* on the length *and* on the amplitude.

One place where the resulting nonlinearity produces dramatic effects is on the seashore. As an incoming ocean wave approaches the beach, the sea suddenly becomes shallow and nonlinear effects make the base of the wave slow up. As a result, the faster-moving wave top overtakes the base, and the wave topples over and "breaks" onto the beach.

Under the circumstances of shallow, nonlinear waves, an interesting possibility arises. If a collection of waves with different amplitudes and wavelengths are superimposed in the right way, then the effect on the wave speed due to dispersion can indeed be exactly compensated by the effect on

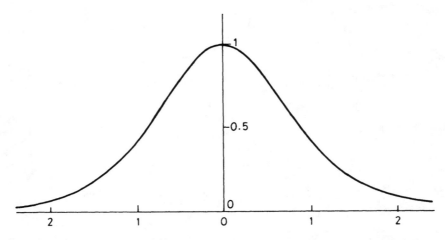

Figure 7. The "soliton" solution to the equation derived by Korteweg and de Vries: this resembles the humps of water observed by John Scott Russell.

the speed of the different amplitudes. This was the essence of what Russell observed. And there is no great trick to achieving the "right" mix of amplitudes and wavelengths, even with a horse-drawn canal boat, because any waves that do not fit the pattern will indeed disperse and fade away, leaving the "right" ones behind. Instead of dispersing, all these remaining contributing waves will propagate at exactly the same speed. The original waveform will thus retain its shape as it moves, even if this shape consists of merely a single hump.

Korteweg and de Vries demonstrated their explanation of solitary waves by proposing an equation to describe the propagation of nonlinear disturbances. They readily found an exact solution to their equation corresponding to a hump of water with the shape seen by Russell, which remains unchanged as it moves (Figure 7). The speed of such a hump depends on its height; big humps travel faster than small humps.

The work of Korteweg and de Vries seemed to provide a satisfactory explanation of the type of phenomenon observed by Russell, but had no other obvious application, and very little further work was done on the topic for seventy years. Solitary waves were regarded merely as a scientific curiosity, of little practical or theoretical importance. And further work was, in any case, impeded by the difficulty of dealing with equations describing nonlinear phenomena. Almost all the techniques devised by mathematicians to treat physical problems over the past three hundred years have been adapted to linear systems; nonlinear systems are notoriously hard to study mathematically.

In the mid-1960s, the advent of computers changed things. People began to investigate the nonlinear Korteweg–de Vries equation using computers to simulate the behavior of solitary waves. In 1965, Martin Kruskal (who had previously carried out important work on the subject of black holes) tried computing the effects of colliding two solitary waves of different height with one another in a "numerical experiment." The results were a great surprise. Intuitively, one would expect the humps, whose stability depends upon a careful balance between nonlinear and dispersive effects, to break up in collision. Instead, the two humps emerged from the amalgamated disturbance intact, and traveled on their way serenely with their original speeds. It is as though each solitary wave has a certain discrete identity, which can even survive an encounter with another solitary wave. So striking were these computer results that Kruskal and his colleagues invented the word *soliton* to describe such humps. The choice of name was motivated by the close analogy with subnuclear particles, such as the proton and neutron, that also have wavelike properties yet retain a discrete identity.

With interest in nonlinear waves rekindled, the subject began to take off. It was soon found that, far from being freak entities, solitons could form in a very wide range of physical systems, not merely on the surface of shallow water channels. The key feature in all cases is indeed nonlinearity. So long as a medium is nonlinear, it is likely to propagate pulses of energy in the form of solitons—or at least, something very similar. The actual medium in which the solitons propagate is irrelevant. It may be a liquid, a solid, a gas, an electric current or a field such as the electromagnetic field. Solitons have been studied in systems as diverse as planetary atmospheres, crystals, plasmas, optical fibers, nerve fibers and electronic devices.

One unexpected field of application is molecular biology. There has been long-standing controversy over the mechanism whereby focused energy can be transported along biological long-chain molecules such as proteins or DNA, leading to reactions a very great distance (in molecular terms) from the site of the energy production. Some biologists now believe that this cannot be due to ordinary chemical processes, but that energy is carried by soliton ripples in the structure of the molecule.

Another area of research in which the concept of solitons has made a big impact is superconductivity, especially so-called high-temperature superconductivity. Some materials become superconducting at temperatures close to absolute zero ($-273\,°C$) because of the way electrons can pair up and move in an organized way in the absence of substantial thermal "noise." Discoveries made in the late 1980s have shown that certain ceramic materials are superconducting at much higher temperatures, and there is speculation that room-temperature superconductors might be on the horizon. If so, the technological applications could prove revolutionary. But how can materials be superconducting at such high temperatures, where we would expect any electron pairs that tried to form to be disrupted by thermal agitation?

Although mystery still surrounds what is happening in the new high-temperature superconductors, theorists suspect that solitons play a key role. Solitons have already been observed in other superconducting devices, the so-called Josephson junctions in which a narrow insulating barrier separates two sections of superconducting material. When a current flows in such a device, discrete packets of magnetic field energy, a form of solitons known as fluxons, tunnel through the barrier quantum-mechanically. Researchers hope that these magnetic field solitons may one day be used to store digital information in ultrahigh-speed computers—and they suspect that related soliton effects can explain the superconductivity of some ceramics at relatively high temperatures. It may work like this.

Apart from fluxons, there is another type of solid-state soliton called the polaron. It is essentially a solitary wave of electric charge. When an electron moves through a solid substance with a crystalline structure, the electric field of the electron attracts the atoms in the crystal lattice, thereby deforming the state of the lattice slightly. For small deformations the atoms behave like perfect elastic—their movement is in direct proportion to the force acting on them. This means that the system is linear, and no solitons can form. In some materials, however, the deformation can be relatively large, and then the amount of movement is no longer simply proportional to the force. This all-important nonlinearity opens the way to the formation of solitons. In this case, the soliton consists of an electron and the surrounding atoms of the crystal lattice binding together to form a concentrated lump of electrical energy that can propagate through the lattice. There is speculation that two such polarons can interact to produce a bound system—a bipolaron—which produces superconductivity at high temperatures in ceramics in the same way that bound pairs of electrons produce superconductivity in other materials at very low temperatures.

Kinks and twists

What are the general principles underlying this bewildering variety of solitons? The characteristic feature of solitons is that they have a type of permanence. But there is an important distinction between truly permanent solitons and those that are merely long-lived. Solitons on the surface of water, for example, could be destroyed, if need be, by making a large disturbance in the water. By contrast, some solitons can *never* be destroyed.

To understand the distinction, imagine an infinitely long strip of stretched elastic, colored blue on one side and red on the other. Waves will propagate along the elastic if it is plucked. If the elastic were nonlinear, a soliton could be produced by raising a hump and releasing it. The soliton would be a lump of concentrated energy. The energy concerned is, in this case, elastic energy associated with the deformation of the strip, an extra stretch, at the hump. Because the hump can in principle be flattened out again, the soliton is not permanent.

But another type of soliton is possible on the strip. This can be produced by a *twist* in the strip, so that to the left of the twist the red side is uppermost, while to the right the blue side is uppermost (Figure 8). Once again, the energy is concentrated in a lump, but this time the soliton cannot be destroyed. Although the twist can be slid back and forth along the strip, there is no way that it can be untwisted, if the elastic strip is infinitely long (indeed, there is no way in

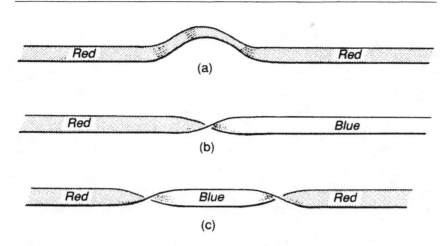

Figure 8. Concentrated regions of elastic energy (solitons) can be produced on an elastic strip in two ways, as either humps or twists. Both can travel along the strip, but a hump (a) could be disrupted and eliminated by a disturbance. By contrast, a twist is "topologically trapped" on the strip (b), and can only be eliminated by encountering an "antitwist" (c).

which such a soliton can be created except when the infinite strip itself is created; it is an integral feature of the twisted strip). There is, though, the possibility that this soliton could encounter an "antisoliton," consisting of a twist in the opposite direction moving along the strip. In that case the two twists would annihilate each other, and their energy would be converted to ordinary waves in the strip. The analogy with a particle annihilating with an antiparticle and releasing energy is very close, and we can even envisage the creation of soliton-antisoliton pairs (though not individual solitons of this kind) by using energy to twist a short section of the strip so that a soliton is created at one end of the twisted section and an antisoliton at the other end.

The investigation of twists is a branch of the science of topology. Topologists study how lines, surfaces and volumes can be twisted, knotted and connected together, if not in reality then at least in simulation, with the aid of mathematics. A system that has a certain topology cannot change it by mere stretching and bending—the distorted shapes that result from such processes are equivalent to one another, topologically speaking. The only way to change the topology of something is by cutting and rejoining. No amount of contortion can untwist a twisted infinite strip, or unknot a closed loop of rope with a knot in it. So solitons that are topologically trapped will endure so long as the system concerned endures.

Such topological solitons appear in many guises. In a crystal, for example, dislocations occur where the regular arrangements of atoms in a lattice are distorted by a mismatch. Although these dislocations can move around within the crystal, they can never be eliminated. Other examples of solitons are found, once again, in superconductors, where a concentrated magnetic field can be trapped in a thin tube. And something very much like this underlies the explanation for cosmic strings—a phenomenon we discuss in Chapter 6.

Perhaps the most promising area for topological solitons, though, lies in subatomic particle physics. Here solitons appear as excitations of fields, rather than of some material medium. When a field is in its lowest energy state, it is uniform throughout space. Excitations arise when the field departs from uniformity somewhere. In a nonlinear field it can happen that, as a result of the interaction of a field with itself, the lowest energy state is no longer a state of zero field. That is, the total energy is lower when the field is present with a finite strength than when it is absent altogether. This is because the effect of the field acting on itself is to reduce its energy. In these cases the field will still be uniform throughout space, but with a nonzero value.

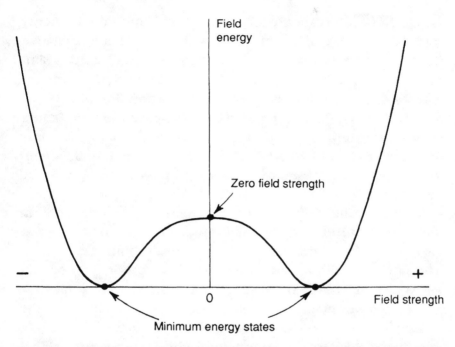

Figure 9. A graph of field *energy* against field *strength* for a typical nonlinear field that arises in subnuclear particle physics. The state of zero field strength possesses *nonzero* energy (the top of the hill). There are two possible states with zero field energy (at the bases of the valleys), one with positive field strength and the other with negative field strength. These are analogous to the two sides of the strip in Figure 8.

But now a new possibility arises. There may be more than one constant value for the field, just as there was for the twisted elastic strip. In this case the two sides of the strip correspond to the field having either positive or negative value.

Figure 9 shows a graph of the energy for a typical nonlinear field. The state of zero field strength corresponds to the top of a symmetrically shaped hill between two valleys. Each

valley floor corresponds to one of the minimum energy
states of nonzero uniform field, with positive and negative
field values, respectively. If the energy of a field has this simple dual-minimum form, then it may happen that in one region of space the field is in a state corresponding to the left-
hand valley, while in another region of space the field is in a
state corresponding to the right-hand valley. If so, the only
way that the field can be continuously joined is if, somewhere between these two regions, it passes through value
zero—that is, it "climbs the hill." Where this occurs there
will be a localized region of field energy. This is the soliton.
Like the twist in the strip, it is topologically trapped between
two physically distinct regions of identical minimum energy.
Though the soliton can move about, it can never be destroyed—unless, of course, it encounters an antisoliton.

The analogy with the elastic strip is limited, because the
soliton on the strip can move only in one dimension (along
the strip). Real fields extend throughout three-dimensional
space. To develop the concept of a topological soliton in
three dimensions requires a more advanced and abstract use
of topology. Nevertheless, the essential idea remains the
same—the field configuration contains a topologically
trapped, localized region of energy which can move about in
space, but cannot "unwind" itself.

Many theorists believe that such solitons would appear to
us as new types of subnuclear particles with rich and interesting properties. Indeed, ordinary protons, neutrons and the
rest of the particle zoo can be regarded, in a certain basic
sense, as solitons in the appropriate force field. One way in
which "new" types of soliton particles might reveal themselves is by displaying properties that are absent in ordinary
particles. A classic example was discovered (mathematically)
in the early 1970s by Gerard t'Hooft of the University of

Utrecht and Alexander Polyakov of the Landau Institute for Theoretical Physics, in Moscow. They were investigating a new type of subnuclear field thought to be responsible for the strong nuclear force when they discovered that this field possesses a multiplicity of minimum energy states between which the field can be "twisted" or "knotted." In some of these configurations the resulting soliton would act like an isolated magnetic "charge." All known magnets have both a north and a south pole, so a single pole would be most distinctive. In spite of extensive searches, however, no experiment has yet found unambiguous, convincing evidence for the existence of such magnetic monopoles.

Recently, the concept of a soliton has been extended from three dimensions to four, by including time as well as space in the calculations. A four-dimensional soliton is an entity that is localized in both space and time, so that it enjoys only a fleeting existence. Nevertheless, such "instantons," as they have become known, can play an important role in the subatomic world by allowing transitions to occur between field configurations in ways that were previously assumed to be forbidden. Roughly speaking, a field can change from one configuration to another by twisting as it goes.

The study of solitons, instantons, kinks and other topological features is finding applications in branches of science as diverse as biophysics and cosmology. It is now widely believed that during the very early stages of the Universe—during the big bang—physical processes were dominated by nonlinear fields. These could have created topological structures that might have survived in the Universe until the present day—one possible example concerns the threadlike entities that go by the name of cosmic strings, which are discussed in Chapter 6.

The growth of nonlinear science has been quite phenomenal in recent years, thanks largely to the availability of fast computers. This burgeoning study of nonlinear systems is causing a remarkable shift of emphasis away from inert "things"—lumpen matter responding to impressed forces—and toward "systems" that contain elements of spontaneity and surprise. The old machine vocabulary of science is giving way to language more reminiscent of biology than physics—adaptation, coherence, organization, and so on. In many cases the same basic nonlinear phenomena are manifesting themselves in systems that are not really material at all, including computer networks and economic models. So with the machine analogy now looking distinctly strained, the link with Newtonian materialism is fading fast. The very breadth of the nonlinear revolution is leading to the rapid demise of the Newtonian paradigm as the basis of our understanding of reality.

In spite of the post-Newtonian flavor of these developments, however, most studies of nonlinear systems still retain Newton's conception of space and time. Even though interest focuses on systems rather than mechanisms, these systems can still be envisaged as inhabiting an absolute space and time. But we have known for nearly a century now that even these elements of Newtonian materialism and the clockwork Universe must also go into the melting pot, with consequences no less profound than those we have already discussed.

3 The Mysterious Present

Albert Einstein taught us that space and time are not what they seem to the human senses. For a start, they should be regarded as two connected facets of a greater whole, called spacetime. From the more holistic viewpoint of relativity theory, concepts such as length, mass and duration take on a much more nebulous aspect than they do in the apparently rigid reality of our everyday world. Even the idea of simultaneity and the concept of "now" assume an elusive character that often runs counter to common sense. What relativity theory takes away with one hand, however, it gives back with the other, in the form of "new" and truly fundamental constants and concepts.

The arena of space

Most people take space for granted. It is such a fundamental part of experience that we are inclined to accept it without question. How could space be otherwise than it is? Common sense suggests that space is so basic to reality that its existence and properties are hardly worthy of contemplation. Doubt only begins to creep in when we are confronted with questions such as: Does space extend forever? Did space exist before the Universe was created? At that point, another question arises: Where did the "common sense" view of space come from in the first place?

Historians trace the origin of the common sense view back at least as far as Ancient Greece, where it was intimately bound up with the development of geometry. Geometry received its most systematic formulation several centuries before Christ, culminating in the work of Euclid.

The geometers, in constructing their theorems, introduced idealized concepts like parallel lines, which were defined to extend infinitely without crossing. The existence of such lines was needed in order to enable the theorems to be proved; they also, though, implicitly required the existence of an infinity "out there" into which lines could, in principle, be projected. All this was innocuous enough so long as the geometers' space remained abstract; but problems arose when the space of geometry began to be identified with physical space, in the real world. Early attempts to do this can be found with the work of the Atomists who, as we mentioned in Chapter 1, postulated (long before Euclid) that the world consists of just two things: indestructible particles (atoms) and a limitless void. The void was envisaged as an arena in which atoms move and the great drama of nature is acted out. This image is very close to most people's common sense notion of space today.

The theory of an infinite void, apparently required for the projection of parallel lines, came into direct conflict with Greek cosmology, which held that the Universe was finite and spherical, with the Earth at the center of a system of concentric revolving spheres. The question of what, if anything, lay beyond the outermost sphere was deeply troubling. Aristotle, in the fourth century B.C., tried to evade the issue by adopting a curious definition of "place," He asserted that the outermost sphere isn't *in* anything; it contains, but is not itself contained. In short, there is no outside.

Supporters of the idea of the void repeatedly countered with variations on the following conundrum. Suppose one were to travel to the farthest limit of the Universe and stretch out one's hand (or throw a spear, to use the example favored by the Roman poet Lucretius). What would be encountered?

More space? A solid wall? Would the hand (spear) fade away, or suddenly cease to exist?

The controversy raged for many centuries, right up to the Renaissance and the rise of the modern scientific era. Under the impact of Copernicus, Galileo and Newton, the ancient idea of a finite spherical world was eventually abandoned, and the Atomist concept of limitless space containing the stars and planets at last became generally accepted. But now a new difficulty arose. Newton conceived of space in more than purely geometrical terms, for he was primarily concerned with the construction of mathematical laws of motion. These required space to have *mechanical* properties, as well.

Absolute space and the laws of motion

One of the oldest problems of science and philosophy is the distinction between absolute and relative motion. A common experience of the latter is the impression that your train has begun to move out of a station, whereas in fact it is an adjacent train that is moving off in the opposite direction. Even more memorably, one of the authors was once traveling on a car ferry which had started to pull out from the harbor so smoothly that a passenger nearby, glancing up from a book she had been reading, cried out "My God: The sky is moving!" By contrast, though, a ride on a roller coaster leaves no doubt that it is you who are moving, because of the physical effects that the accelerations have on your stomach. Acceleration is clearly different from uniform motion.

Newton's famous laws of motion embody what is now known as the principle of relativity, discovered (or formulated) earlier by Galileo. The principle is best described by an example. Imagine that you are aboard an aircraft flying straight and level at fixed speed. From within the aircraft there is no sensation of motion whatever. Activities like pouring a drink or walking about seem perfectly normal. According to Galileo and Newton, this is because uniform motion in a straight line is purely *relative;* that is, it has meaning only when referred to some other object or system. Thus, to say that an object has a definite speed is meaningless. Its speed must be specified relative to something else. When we say that a car is traveling at 30 miles per hour, we really mean 30 mph *relative to the road.* The distinction becomes important if the car is in collision with another car traveling the opposite way at 30 mph; the relative speed of the two cars is 60 mph, not 30 mph, and the resulting damage is commensurate with the higher speed. In particular, though, one must relinquish any idea of an object having a definite speed *through space.* Space, being empty, provides no landmarks against which, say, the speed of the Earth can be judged. Deciding the speed of the Earth depends on what you are measuring relative to—the Moon, the Sun, the planet Jupiter or the center of the Milky Way Galaxy? Likewise, no particular object can be said to be in a state of absolute rest in space. *Star Trek* stories that have the *USS Enterprise* "stopped dead in space" due to power failure appeal to pre-Renaissance physics.

In the case of uniform motion in a straight line there is no distinction, then, between real and apparent motion. Things are very different, however, when it comes to nonuniform motion. If the aircraft you are traveling in banks steeply into a turn, or simply changes speed sharply, the effects can be clearly felt as forces that pull or push at your body, and activities like pouring a drink or walking become much harder.

Newton explained that such effects are caused by "inertia." Although objects have no resistance to uniform motion in free space, every object possesses a natural resistance to *changes* in motion. These may be accelerations in a straight line or alterations in the direction of motion, or both; the object tries to keep on going as it was, in the face of the changes. The most familiar examples of inertia at work, and ones that Newton was especially interested in, involve rotating objects, which experience centrifugal forces. Anyone who has ridden a carousel, or been in a car taking a corner at high speed, has experienced centrifugal forces at work.

This contrast between uniform and nonuniform motion is deep. Whereas uniform motion is relative, nonuniform motion seems to be absolute; one can certainly say that an object is accelerating without direct reference to anything external. Thus, the carousel riders know they are rotating without having to look out at the fairground whirling by; they can tell this with their eyes shut, and they are quite sure that it is the carousel that moves, not the surroundings (or the sky!). Newton himself came to the conclusion that this kind of motion that seems to require no reference to other objects must be referred to space itself. He invented the concept of "absolute space," regarding it in some respects as like a substance enveloping all objects, and within which objects could be said to accelerate. According to this view, it is the reaction of absolute space back on an accelerating object—a kind of dragging effect, like pushing your hand through water—that produces inertia, or centrifugal force.

In developing this idea Newton conceived of a thought experiment[1] along the following lines. Imagine a bucket of water suspended from a long rope so that the bucket is free to spin. Suppose that the rope is strongly twisted and then released, so that as it unwinds, the bucket rotates (Figure 10). At first, the water remains unaffected. Then, as the rotation of the bucket begins to influence the water inside by viscous drag, eventually the water and the bucket will be ro-

1 That is, one which is carried out only in the imagination.

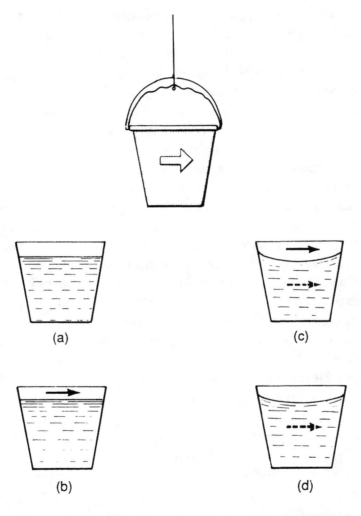

(a)   (c)

(b)   (d)

Figure 10. Newton's bucket experiment. The rope is twisted and the bucket of water released. Before the experiment, the water is at rest relative to the bucket, and the situation is as shown in *a*, with the water surface flat. When the bucket starts to rotate (solid arrow), the water surface remains flat, as in *b*. Eventually the water corotates with the bucket (broken arrow) and the surface becomes concave, as in *c*. If the bucket is stopped but the water swirls on, as in *d*, the surface remains concave. Evidently, the state of the water surface is *not* connected with its motion relative to the bucket.

tating at the same speed. When the water rotates, its surface will adopt a concave shape, piling up against the rim of the bucket because of centrifugal force; and if you now catch hold of the bucket and stop its rotation, for a time the water will continue to swirl around inside, maintaining the concave shape.

You can tell that the water is rotating simply by looking at the shape of its surface. No reference to anything else in the Universe is necessary. The water is nonrotating when the surface is flat, and it is rotating when the surface is concave. In particular, the concavity does *not* depend on motion relative to the bucket that holds the water. In the first part of the experiment, the bucket rotates relative to the water, but the surface of the water is flat. At the end of the experiment, the water rotates relative to the bucket, and the surface is still concave. In the middle of the experiment, there is no relative motion between the water and the bucket, but the surface is still concave—whereas, before the experiment even began, there was no relative movement between the water and the bucket, but the surface was flat. The curvature seems to depend upon the *absolute* rotation of the water—rotation relative to what Newton called absolute space.

You can take the thought experiment a little farther by imagining that it is carried out at the North Pole. Now, even when the bucket has stopped rotating and the water has stopped swirling around in the bucket, careful measurements will still show a slight concavity in the water surface. This is because the rotation of the Earth is carrying the water around with it—that same rotation of the Earth that, for the same reason (centrifugal force) makes the planet bulge outward at the equator. Rotation is not something that should be referred to the Earth either, or indeed to the Sun, the planet Jupiter or the center of the Milky Way Galaxy. The water surface will, in fact, be flat only when it remains stationary (nonrotating) with respect to the most distant concentrations of matter in the Universe, far-off galaxies and quasars.

Now, according to Newton the water surface is flat when it does not rotate relative to absolute space. So the frame of reference that defines absolute space seems to be the same as the frame of reference in which distant galaxies are located. This is the same as saying that the whole assemblage of galaxies is not rotating—that the Universe as a whole does not rotate, even though every known system within it, including planets, stars and individual galaxies, does rotate. The logic of this argument appeals to our common sense, perhaps because our common sense is based on three hundred years of Newtonian physics. But there is an alternative view.

A contemporary rival of Newton's, Gottfried Leibniz, proclaimed that "There is no space where there is no matter." Some years later, the philosopher Bishop George Berkeley also denounced the idea of absolute space as meaningless. "It suffices to replace absolute space," opined Berkeley, "by a relative space determined by the heaven of fixed stars."[2] Regarding nonuniform motion, Berkeley wrote: "I believe we may find all the absolute motion we can frame an idea of, to be at bottom no other than relative motion." Berkeley considered that all motion, including acceleration and rotation, should be regarded as relative to the fixed "stars," and not to space itself.

In spelling out his argument, Berkeley asked his readers to envisage a spherical object ("a globe") in an otherwise totally empty universe. In this featureless void, argued Berkeley, no motion of the sphere can be conceived. Not just steady

2 He referred to "stars," rather than "galaxies," because galaxies beyond the Milky Way had not, at that time, been identified.

movement through space, but acceleration and rotation as well, are meaningless. Now imagine a universe that contains just two spheres joined by a rope. It is possible now to imagine a relative motion along the line between the two spheres, but "a circular motion of the two globes around a common centre cannot be conceived by the imagination." On the other hand, "let us suppose that the sky of fixed stars is created." Then the rotation *can* be perceived, against this backdrop.

This flatly opposes Newton's view of what would happen in Berkeley's hypothetical universe. Even a solitary globe would be deemed to be rotating if its equator bulged outward; and the rotation about a common center of two globes joined by a rope could be determined by measuring the tension in the rope produced by the appropriate centrifugal force. Newton explicitly pointed out that "The effects which distinguish absolute from relative motion are centrifugal forces. . . . For in a circular motion which is purely relative no such forces exist."

In spite of the sweeping success of Newton's mechanics and the world view it engendered, the tricky issue of absolute space and absolute rotation did not go away. In the second half of the nineteenth century these matters were taken up by the Austrian physicist and philosopher Ernst Mach, best known for his work on sound waves and immortalized in the concept of Mach number, a measure of speed in terms of the speed of sound. Mach refused to entertain the notion of an unobservable absolute space, and like Berkeley he asserted that both uniform and nonuniform motion were entirely relative. Rotation, for example, is relative to the "fixed stars." But this still left the problem of centrifugal force. If it wasn't caused by the dragging effect of absolute space, where did it come from? Mach proposed a neat solution. From the viewpoint of a rotating observer, centrifugal force is felt whenever the stars are seen to be whirling around. Clearly, asserted Mach, the stars *cause* the force. That is, centrifugal force—more generally, the inertia of an object—has its origin not in some mysterious absolute space enveloping the object, but in the material objects in the far-flung regions of the cosmos. The idea, which became known as Mach's principle, asserts, in short, that the stomach-churning effect of a roller-coaster ride is caused by the distant stars (galaxies) pulling on the organs of your body.

Although Mach failed to provide a very clear formulation of how this might work, the idea that inertia and inertial forces are somehow produced by an interaction between an object and the distant matter in the Universe (Mach's principle) had a profound effect on later thinkers. Einstein, for example, acknowledged that Mach's book *The Science of Mechanics* strongly influenced him in the construction of his own theory of gravity, called the general theory of relativity.[3] By then, however, Einstein had already overturned established ideas about the nature of space and time with his special theory of relativity, published in 1905.

Einstein's insight

Newton's laws, applied to uniform motion in which the speed and direction of travel of different objects remain constant, are the same for all uniformly moving observers, and they deny to any observer or material object the privilege of defining a standard of absolute rest. In this context, the question of the speed of the Earth through space is meaningless, just as the *USS Enterprise* cannot be "stopped dead" in space. But in the mid-nineteenth century the question of the Earth's

3 Ironically, Mach rejected the general theory of relativity, published in 1915, and at the time of his death the following year (the day after his seventy-eighth birthday) he was planning to write a book refuting Einstein's ideas.

speed through space received a new twist. The work of Michael Faraday and James Clerk Maxwell, in particular, revealed the existence of the electromagnetic field as the agency responsible for transmitting electric and magnetic forces across seemingly empty space, and Maxwell derived the equations which now bear his name, and which describe the way waves of electromagnetic force wiggle their way through space. Maxwell was able to calculate the speed of these waves, from his equations, and found that it is exactly the speed of light: 300,000 kilometers a second. Since the speed of light was already known, but the nature of light was unclear, this was a key piece of evidence in establishing that light is a form of electromagnetic wave (we now know that radio waves, x-rays and many other forms of radiation are also electromagnetic waves, and that they all travel at the same speed). But the curious thing about this number that emerged from Maxwell's equations—the speed of light—was that it is a fixed number determined by the equations alone. Where, physicists wondered, is the reference frame relative to which the speed is to be measured? This was how the idea of the ether came to be conceived—as a mysterious jellylike medium filling all space. Electromagnetic waves, being now thought of as vibrations traveling through the ether, must have their speed measured relative to the ether. And this immediately suggested that there must be an absolute sense in which the motion of the Earth could be measured, not relative to empty space, but relative to the ether.

The presence of an ether would define a frame of reference for the state of absolute rest, against which the motion of all material objects could be judged. So in the last two decades of the nineteenth century one of the main focuses of attention among physicists was the effort to measure the motion of the Earth through the ether. If light traveled at a fixed speed through the ether, then it ought to be possible to measure differences in the speed of light by looking in the direction of the Earth's motion (when the Earth is running head-on into a beam of light) and at right angles to the Earth's motion. It came as a bombshell when the appropriate experiments, most notably carried out by Albert Michelson and Edward Morley, in the United States, showed that the speed of light is the same in all directions. They found no evidence whatever of an influence caused by the Earth's motion through the ether.

Although it was Albert Einstein who grasped the nettle and, a few years later, published a new theory—his special theory of relativity—which explained the absence of "ether drift," several other scientists were looking at this puzzle at the end of the nineteenth century, and there is little doubt that the special theory of relativity was an idea whose time had come, and that it would soon have been developed even without Einstein's genius. The key feature of the theory is, however, genuinely revolutionary. It proposes that the ether does not exist, and that the reason why Maxwell's equations give a unique value for the speed of light is that this is a genuine universal constant—that the speed of light has the same value *irrespective of the state of motion of whoever measures it*. Furthermore, this unique constant, *the* speed of light, defines an absolute upper speed limit for all relative motion between material objects—nobody, in any frame of reference, will ever measure the motion of another material object and find that it is traveling faster than light.

All of the strangeness of special relativity, such as the well-known contraction of moving objects and the dilation of time, stems from this fact, the universal constancy of the speed of light. We can give an indication of what is involved by describing another simple thought experiment. Imagine a moving train in which one carriage contains a source of light,

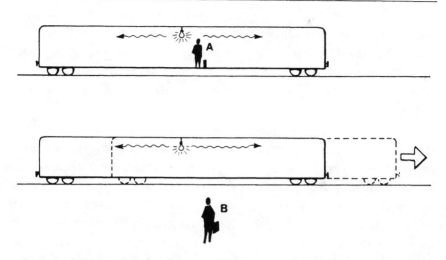

Figure 11. The elusive "now." A lamp at the center of a railway carriage flashes light pulses toward the carriage ends. Everyone agrees that the pulses start out at the same moment. But do they arrive at the ends of the carriage at the same moment?

(i) From the reference point of observer A, traveling with the train, the pulses travel at equal speeds for equal distances, so their arrival "events" are judged to be simultaneous.

(ii) Viewed from the reference frame of the trackside, observer B also sees the pulses travel at the same speed, but the distances are not equal—during their flight, the carriage moves forward, shortening the journey for one pulse and increasing it for the other. B sees the left pulse arrive first. The conflict arises because both A and B see light always travel at the *same* speed.

set up exactly in the middle of the carriage. At a certain moment, two light pulses are emitted in opposite directions, toward the front and back of the carriage (Figure 11). An observer riding in the train will regard the train as at rest relative to herself, and will therefore deduce that both pulses arrive at the end walls of the carriage at the same moment, since they each travel at the same speed and each have the same distance to cover.

Now envisage these events as seen by another observer,

standing on a station platform as the train rushes through. According to Einstein's basic postulate, the speed of light is constant for this observer too—it is the same for each pulse, and the same as the speed of light measured by the observer in the train. From the standpoint of the observer on the platform, though, the train is definitely moving, so the observer sees the rear end of the carriage advancing toward the light pulse, while the front end is retreating from its counterpart. In the time taken for the pulses to reach their respective ends of the carriage, the train has moved a certain distance. So the rearward-traveling pulse has less distance to cover than the forward-traveling pulse; and since both travel at the same speed, that means that the observer on the platform will experience the rearward-traveling pulse arriving at the end of the carriage first.

What can we conclude from this thought experiment? Comparing the same set of events as witnessed by the two observers, a pair of events (the arrival of the light pulses at the carriage ends) that are reckoned to be simultaneous by one observer are perceived to occur at different times by another observer who is moving differently. In other words, the simultaneity of events that are separated in space is *relative*. Different observers in different states of motion measure different durations between the same pair of events.

In a similar fashion, it turns out that different observers in different states of motion will measure different *distances* between the same pair of events. We do not intend to go into the mathematical details here,[4] but just as it is no longer possible to talk of *the* time interval between two spatially separated events, such as the arrival of two pulses of light at op-

---

4 An excellent introduction can be found in Clifford Will's *Was Einstein Right?*.

posite ends of a railway carriage, so it is no longer possible to talk of *the* distance between two spatially separated objects. By traveling in a spaceship close to the speed of light, it would be possible, for example, to see the distance between the Earth and the Sun reduced from the 150 million kilometers we measure to, say, a mere 15 kilometers.

The marriage of space and time

Both space and time, individually, lose their independent status in Einstein's theory. But the combination of the two, spacetime, assumes a truly fundamental significance that is not apparent when we consider either of the two components separately. When an observer changes his or her state of motion, the relationship between space and time is altered, so that distances and durations are perceived differently. But because space and time are separate facets of a greater whole, spacetime itself retains some constant features even for observers moving in different ways. Although time remains physically distinct from space, time and three-dimensional space are so intimately bound together in their properties that it makes sense to describe them jointly, as a four-dimensional "continuum," using mathematical language that takes account of the physical distinction.

The idea can be understood by analogy with ordinary three-dimensional space. Imagine viewing a broom handle from various orientations. The apparent length of the broom handle will vary according to the angles involved (Figure 12). If it is viewed square-on it will present its true length, but when it is oriented obliquely its length will be foreshortened. In the extreme case of being viewed end-on it will seem to have no length at all. Now the human brain has evolved to make sense of this situation. We accept that the length of the broom handle is really fixed and that the varia-

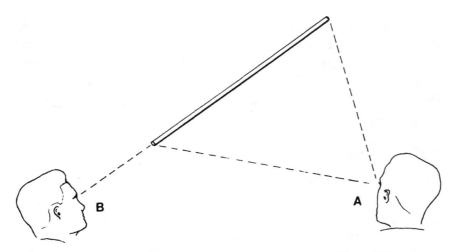

Figure 12. The apparent length of a rod depends on the angle of view. Broadside on, observer A sees the maximum length; edge-on, observer B sees the length foreshortened to zero.

tions in the apparent length are due to the fact that it can lie at different orientations in three-dimensional space.

In fact, there is a simple mathematical formula that relates the true length of the broom handle to the apparent lengths presented to observers in the three perpendicular dimensions of space. The formula says: "To get the true length, square each of the three perpendicular apparent lengths, add these three squares together, and take the square root" (Figure 13). Readers may recognize this as a generalization of Pythagoras' famous theorem involving right-angle triangles. The human brain evidently achieves this computational feat without explicitly performing any mathematical calculations, and we regard the result as intuitively obvious.

In the case of four-dimensional spacetime we must think of an object such as a broom handle having a four-dimensional length. What does this mean? It means we must also take into account the instants of time at which we observe

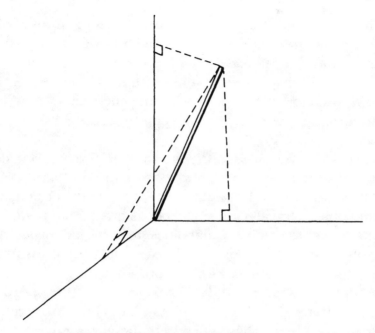

Figure 13. The true length of the rod can be computed from three projected lengths along three perpendicular axes, by a generalization of Pythagoras' theorem.

the respective ends of the handle. If these observation events are at different times, the broom handle will have temporal as well as spatial extension. In this four-dimensional situation, there is an appropriate variation of apparent four-dimensional length depending on the angle of view. Now, because we are dealing with four dimensions rather than three, there is a greater range of orientations available. We know how to vary our orientation in space, but how does one vary orientation between, say, the vertical direction of space, and the direction of time? The answer is by *moving* in the vertical direction. To achieve a noticeable effect, this motion must be at a sizable fraction of the speed of light. The effect is once again to alter the apparent length of the broom handle, making it appear shorter in the direction of motion. This is the length-contraction effect we mentioned before. Conversely, time intervals are stretched, or dilated by the motion. In a sense, an interval of space is traded for an interval of time. So how much space is each unit of time worth? Since the conversion factor is the speed of light, one second is worth the distance light travels in one second—about 300,000 km, or one light-second.

The reason we don't see the world as four-dimensional is that significant trading between space and time takes place only at near the speed of light, and since we never perceive material objects traveling at such speeds the human brain has had no need to evolve capabilities that would make this trade-off intuitively obvious, in the way that we instinctively understand the foreshortening of a broom handle.

To take a specific example, at 90 percent of the speed of light lengths are shrunk by more than half, and clocks run at less than half speed. These effects are, however, entirely relative to the observers concerned. A "superwitch" traveling with the broom handle at such a speed relative to the ground would notice nothing unusual about either its length or the rate of passage of time. For such an observer it would be the objects that are fixed to the ground that are contracted, with clocks on the ground seeming to run slow compared with a clock attached to the flying broomstick. Thus when observers are in relative motion each sees the *other*'s length contracted and the *other*'s clock running slow.

In spite of the intimate interweaving of space and time into a four-dimensional spacetime, space remains space and time remains time. Mathematically, the distinction is expressed by a slight modification of the Pythagorean formula for combining intervals: the square of the time interval is *subtracted*, rather than added (as well as the time interval being multiplied by the speed of light). This difference leads to some curious consequences. Because both negative and positive quantities are involved in computing the square of the four-dimensional interval, this quantity may be either positive, negative or zero. In the original three-dimensional version of the foreshortened broom handle calculation, one always adds, never subtracts, and the numbers involved, being squares, are themselves always positive, so only positive numbers are produced. In the four-dimensional version, the situation is more complicated.

Suppose, for example, that the two events that mark the extremities of our four-dimensional interval are the explosions of two stars located two light-years apart as measured in the reference frame of the Earth. If the Earth observer reckons the explosions to occur, say, one year apart in time, then the spatial separation (two light-years) outweighs the time separation (one year). Squaring these numbers and subtracting the time part from the space part thus gives $4 - 1$ square light-years. The number 3 is positive, and we conclude that the four-dimensional spacetime interval between the two events is largely spatial in character (that it is spacelike). If, however, the explosions were observed from Earth to occur three years apart rather than one, we would have to subtract the square of 3, which is 9, from 4, giving the result -5 square light-years. This would indicate that the time interval outweighs the space part, so that the spacetime interval is timelike. Readers familiar with complex numbers will recognize that the square of a number can only be negative if the number is imaginary. We shall return to this point in due course.

It may also happen that the space and time parts of a spacetime interval are equal: this would be the case if the stars are located two light-years apart and the explosions occur two years apart. In this case the spacetime interval between the explosion events is $4 - 4 = 0$. The events are no distance apart at all in four dimensions! Such a four-dimensional separation (or lack of separation) is said to be light-like, because the situation here is such that a pulse of light from the first explosion will just reach the second star when it explodes. So the points along the spacetime path of a light pulse can be regarded as having zero four-dimensional separation. Thus, although the path of a light pulse is extended in both space and time, as far as spacetime is concerned there is no distance at all involved. This is sometimes expressed loosely by saying that a photon (a particle of light) visits all the points along its path at the same moment, or that, to a photon, it is no distance at all to cross the Universe.

This unified four-dimensional spacetime description has proved highly successful in explaining many physical phenomena, and is now the accepted view of the physical world. Powerful though it is, it has removed from the picture any vestige of a personal "now," or the division of time into past, present and future. Einstein once expressed this point in a letter to a friend regarding the subject of death. "To us who are committed physicists," he wrote, "the past, present and future are only illusions, however persistent." The reason for this is that, according to relativity theory, time does not "happen" bit by bit, or moment by moment: it is stretched out, like space, in its entirety. Time is simply "there."

To understand why this is so, you must first appreciate that your now and my now are not necessarily the same. This is because, as we have seen, the simultaneity of two spatially separated events is entirely relative. What one observer regards as happening at "the same moment" but at another place, a second observer, located elsewhere, may regard as happening before, or after, that moment. We fail to notice this in everyday life because the speed of light is so great that the time discrepancies involved are minute over Earth distances. On an astronomical scale, however, the effect is enormous. An event in a distant galaxy which we judge to be simultaneous with noon today in a laboratory on Earth can be shifted by centuries, from your point of view, if you happen to change your reference frame by boarding a train.

These ideas have a profound implication. If the "present moment" elsewhere in the Universe depends on how you are moving, a whole span of "presents" must exist, some of which will lie in what you regard as your past, some in your future, as seen by different observers (Figure 14). In other words, moments of time cannot be things which "happen" everywhere at once, in which only the unique present is "real." Rather, time is extended in some way, like space; which particular distant event any given observer regards as happening in the mysterious moment of "now" is purely relative.

So does the future, in some sense, already exist "out there"? Might we be able to foresee events in our own future by changing our state of motion? Indeed, thinking again about the train experiment, if the events we described earlier had been observed from the viewpoint of a passenger in an express overtaking the first train, the time order of the arrival of the two pulses of light at their respective ends of the carriage would have been the reverse of the order seen by the observer on the platform. This seems like "time running backward," in a sense. But it turns out that you cannot travel fast enough to see into your own future. To accomplish that, the information about your future would have to be transmitted so fast that subtracting the time component from the space component of spacetime would leave a negative answer. As we have mentioned, traveling at the speed of light shrinks the four-dimensional interval to zero. Making it smaller still requires travel faster than light, so that the four-dimensional intervals between events become negative. But,

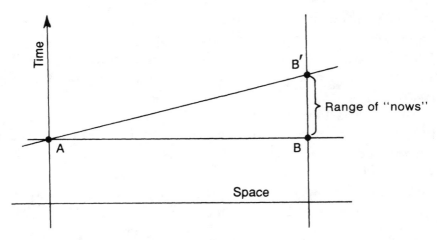

Figure 14. In one reference frame, events B and A are simultaneous: event B occurs at "the same moment" as event A. In another frame, it is event B' that is
simultaneous with A. If A is somebody's "now," which event, B or B', can
uniquely be described as happening "now"? The answer is, *neither.* There is a
whole range of "present moments" including B and B', and any definition of
"now" is entirely relative. By changing one's state of motion, the choice of
simultaneous events can be altered, perhaps by hundreds of years! Any attempt to
argue that only "present moments" are real therefore seems doomed: time must
be stretched out, like space, so that past, present and future exist with equal status.

as we have indicated, that is ruled out by the special theory
of relativity.

More specifically, the theory forbids any physical influence, force or signal to accelerate to a speed faster than light.
This means that only events which have no causal influence
on each other can have their time sequence reversed. In the
case of the train experiment, for example, whatever the reference frame of the observer, the pulses of light always arrive
at the carriage ends *after* they are emitted, never before, because these events are causally connected. Their moments of
arrival *relative to each other* can vary, however, because the

pulses have no further causal influence on each other after they are emitted. As far as cause and effect are concerned, the best we can ever do, by approaching the speed of light itself, is to see two causally connected events occur almost simultaneously; we can never reverse their order, and that applies to all causal sequences. We *can* see an illusion of time reversal, but only in events that do not influence each other; we cannot *cause* things to happen "backward in time."

It is perhaps worth mentioning, briefly, that all of the implications of special relativity, including length contraction, time dilation and the need to measure intervals in the appropriate four-dimensional sense, have been confirmed by direct experiments. There are still people who believe that it is all "just a theory" and dismiss it on the grounds that it clashes with common sense; but they are wrong. Experiments involving subatomic particles, moving at close to the speed of light in accelerators like those of the European CERN laboratory, explicity reveal the effects predicted by Einstein. In many cases these effects are dramatic. For example, the lifetime of an unstable subatomic particle can be observed to be dilated by a factor of twenty or more. In one type of accelerator machine, known as a synchrotron, this slowing of time is put to practical use. The electrons which whirl around inside the synchrotron produce intense beams of electromagnetic radiation which can be used, among other things, to probe defects in metals. Because of time dilation, the frequency of the radiation is drastically less (it has a longer wavelength) than it would appear to an observer moving with a circulating electron. This makes the radiation of much greater practical utility. It is also worth noting that in heavy atoms some of the electrons also orbit at close to the speed of light and are therefore subject to strong relativistic effects.

Sometimes, this influences the overall properties of the material at large. For example, the color of the metal gold is explained in this way—most metals are silvery.

As a result of many decades of careful tests, there is no doubt whatever about the accuracy of the special theory of relativity as a description of space and time from the viewpoint of observers moving at constant velocities relative to each other. The main limitation of the theory is that it cannot deal adequately with nonuniform motions and gravitational fields; that deficiency, however, is exactly what Einstein rectified with his general theory of relativity, which gets its name because it can, indeed, deal with these more general situations.

Getting to grips with gravity

Unlike special relativity, the general theory would probably not have been formulated for many decades after the special theory if it had not been for the personal genius of Albert Einstein. Although a few people, like Mach, worried about the problem of inertia, in the first decades of the twentieth century there were no experiments that showed pressing shortcomings in the special theory (no counterpart to the Michelson-Morley experiment, which showed the pressing shortcomings of Newtonian theory). Einstein developed his masterwork purely as a mathematical description of the Universe—an example of abstract theorizing of the highest order. Apart from some minor observational consequences that were tested soon after the theory was published, it took sixty years, until the discovery of quasars, pulsars and black holes, for Einstein's general theory to come into its own as a branch of applied science, explaining many important features of the Universe we observe. The reason for its widespread application in the astrophysical realm is that all those exotic astronomical objects are associated with intense gravitational fields, and the general theory of relativity is, foremost, a theory of gravity.

Einstein gained his unique insight into the nature of gravitation from puzzling over the origin of the forces associated with nonuniform motion—the inertial forces. He used to say that the flash of inspriation that set him on the path to general relativity came when he realized that a man falling from a roof—or a person trapped in a falling elevator—does not feel the force of gravity. If the acceleration of the falling elevator, plunging downward at an ever increasing speed, can *precisely* cancel the force of gravity, thereby producing weightlessness, then the gravitational force and the inertial force produced by the acceleration are equivalent to one another.

The equivalence between gravitational and accelerative effects is central to Einstein's theory and he elevated it to the status of a fundamental principle. It leads directly to one of the most distinctive predictions of the theory. Imagine being in that falling elevator and watching the path of a light pulse crossing the elevator. In the reference frame of the falling observer the light travels in a straight line; but this means that from the point of view of an observer standing on the ground the light path must curve downward (Figure 15). The latter observer would attribute the curvature of the light beam to the effect of gravity, so Einstein made the prediction that gravity bends light. The prediction was directly tested by the astronomer Arthur Eddington during the 1919 Solar eclipse. Eddington measured a slight displacement in the positions of stars along lines of sight close to the eclipsed face of the Sun; this is attributed to the bending of the starbeams by the Sun's gravity (Figure 16). A more accurate test of the same kind can nowadays be performed by bouncing radar

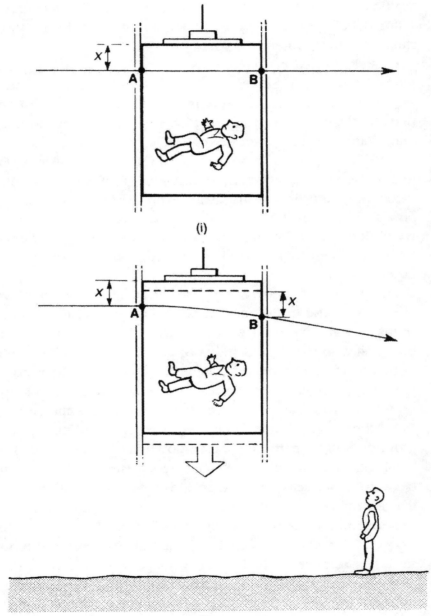

(i)

(ii)

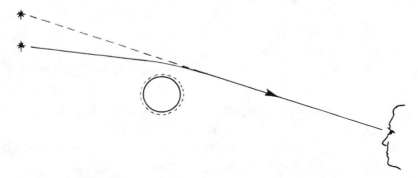

Figure 16. The Sun's gravity bends starbeams, so the position of a star in the sky when seen near the Sun (which is possible during an eclipse) is displaced from its "true" position.

beams off the inner planets of the Solar System, and observing that the echoes are slightly delayed due to the curvature of the paths they follow near the Sun.

The fact that an observer in free fall is weightless makes it seem as though gravity can simply be transformed away by a change of reference frame. But this is not so. Even in a falling elevator an observer could tell that the Earth was exerting a gravitational pull. Objects near the floor of the elevator are slightly closer to the Earth than those near the ceiling.

Figure 15. A photon (light pulse) crosses the interior of a falling elevator, passing through two holes in its sides.

(i) In the reference frame of the hapless occupant (for whom the elevator's reference frame is at rest) the photon enters at A and exits at B, maintaining always a fixed distance x from the roof. Its path appears to be a straight line.

(ii) Viewed from the ground, the elevator accelerates downward during the time it takes the light pulse to cross from A to B. In order to exit from the elevator the same distance below the roof that it entered, the photon must also have fallen, by the same distance. So gravity must bend light rays.

Because the Earth's gravity weakens with distance, the lower objects accelerate slightly faster than the higher ones. So there is a slight tendency for freely falling objects at different heights (whether they are inside an elevator or not) to drift farther apart. In fact, these differential motions are responsible for the tides raised by the Moon's gravity in the oceans on Earth; they are referred to as "tidal forces."

Einstein realized that tidal forces cannot be transformed away by changing the reference frame—they represent a genuine effect of the gravitational field at work. He reasoned that if the effect of these forces is to stretch or distort the distances between freely falling objects, then the most satisfactory description of tidal gravitation is as a distortion or stretching of spacetime itself. That is, rather than regard gravity as a force, Einstein proposed that we regard it as a curvature or warping of spacetime.

In a sense, the curvature of starbeams passing near the Sun can be regarded as a direct probe of the curvature of space. But it is important to appreciate that the curvature involves space*time,* not just space. The Earth follows a closed, elliptical orbit around the Sun, and on first acquaintance with general relativity it is natural to guess that this means the planet is following a path through curved space dictated by the gravitational field of the Sun. But since the Earth's orbit is a closed path, that seems to mean that space is somehow folded completely around the Sun, engulfing the Solar System in what is known as a black hole. Clearly, such an image is badly wrong. The mistake is subtle, but crucial. Viewed in *spacetime,* the Earth's orbit is not a closed ellipse, but a shape like a coiled spring, or helix (Figure 17). After each orbit of the Sun, the Earth returns to the same *place,* but to a different *time,* advancing one year along the time "axis" for each orbit around the Sun. Whenever we consider the time

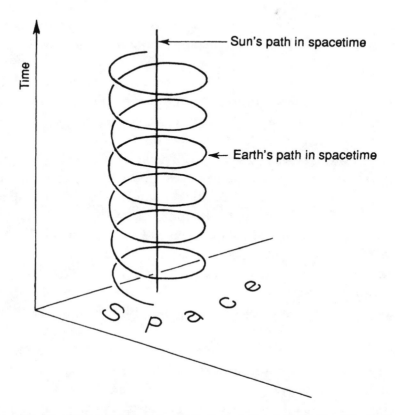

Figure 17. Viewed in spacetime, the Earth traces out a helix as it orbits the Sun. Because intervals of time must be multiplied by the speed of light (a very large number) in order to compare them with distances in space, the helix is actually vastly more elongated vertically than we have shown it.

part of spacetime, we always have to multiply the appropriate numbers by the speed of light, which is a very big number, and this has the effect of stretching out the helix enormously. The "distance" along the time axis that corresponds to a single orbit of the Earth around the Sun is, therefore, a light-year, some 9,500 billion kilometers. So the correct image of the Earth's orbit in terms of curved spacetime is a very shallow curve, weaving around the line that represents the path of the Sun through spacetime. The shallowness of the curve is a consequence of the fact that the Sun's gravity, enormous though it may be by terrestrial standards, is too feeble to cause any but the tiniest spacetime distortions. Later we shall see how in some astronomical objects truly dramatic spacetime curvature effects can occur.

The boldness of Einstein's approach to the puzzle of gravity and nonuniform motions was the abandonment of the idea of flat space, and the introduction of a curved spacetime. Having demolished Newton's mechanics with his special theory, in 1915 Einstein abolished Euclidean geometry as a description of space with his general theory.

But what does curved space, let alone curved spacetime, really mean? Look again at the key feature of Euclidean geometry, the concept of exactly parallel lines that *never* intersect. In the nineteenth century, the mathematicians Karl Gauss, Georg Riemann and Nikolai Ivanovich Lobachevsky had constructed non-Euclidean systems of geometry, in which parallel lines do not exist.. Such geometries are applicable if we want to study curved surfaces; for example, on the surface of the Earth, lines that seem to start out parallel to each other may eventually intersect (Figure 18). The geometry of curved surfaces can therefore have some strange properties, utterly at odds with the theorems of Euclidean geometry that we learn at school. To give an example, a triangle drawn on the surface of a sphere may contain *three* right angles (Figure 19).

When he described gravity in terms of *curved* spacetime, then, Einstein was proposing that non-Euclidean geometry should be applied to spacetime itself. The idea that space and time can be distorted by *motion* was extended to include the influence of *gravity,* so that the presence of matter

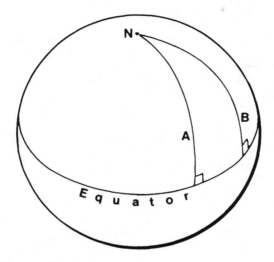

Figure 18. At the Earth's equator, lines of longitude are parallel. Nevertheless, they meet at the poles. This is because the Earth's surface is curved.

in spacetime would also cause distortions, or warpings, of space and time. In Einstein's theory, unlike that of Newton, spacetime must be treated as a mechanical system in its own right. Spacetime is no longer merely an arena in which nature's drama is played out; it is one of the players. This means that there are laws of mechanics for spacetime itself, laws that determine how it can change. And as gravitating objects move about, so the space and time warps that they produce must also change. It is even possible to set up ripples in spacetime, gravitational waves that travel with the speed of light—a phenomenon we shall describe in detail in Chapter 6.

The general theory of relativity provides us with an accurate description of how material bodies move in the presence of gravitational fields, in terms of the curvature of spacetime. John Wheeler, one of the leading physicists involved in the

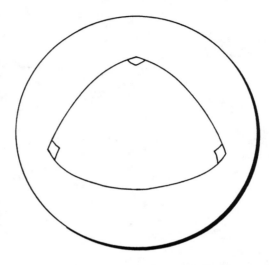

Figure 19. On a spherical surface, the angles of a triangle can add up to 270°—three right angles.

renaissance of general relativity that occurred in the 1960s, expresses this link with the dictum: "Matter tells space how to bend; space tells matter how to move." But general relativity does not successfully incorporate Mach's principle into the picture. The only force that could explain how the influence of distant galaxies might account for the feeling in your stomach when you ride a roller coaster is gravity; but gravity seems at first sight hopelessly feeble for such a task. Newton's famous inverse square law of diminishing gravitational returns still applies within the framework of general relativity, implying that here on Earth the effect of the gravitational force of the entire great galaxy in Andromeda, for example, is about one-hundred-billionth of that produced by the Sun. On the other hand, the density of matter in the Universe is more or less constant on a large scale, so that the amount of matter contained in a spherical shell of space of given thickness centered on the Earth is proportional to the volume of that shell—which is, in turn, proportional to the square of the distance to the shell. So although the gravity of each bit of distant matter affects us only weakly, there is more of the distant matter, and the two effects exactly compensate.

This is an intriguing coincidence, and encourages speculation that when an object starts to rotate it sends out gravitational disturbances into the depths of space, causing all the galaxies in the Universe to move and to react back in unison on the rotating object to produce the observed centrifugal force. Unfortunately, though, such a simple scheme won't work. The reaction of the Universe on the rotating object needs to be instantaneous, but the theory of relativity forbids any physical effect operating faster than light. Even at the speed of light, it would take billions of years for the more distant galaxies to produce a response on Earth. Any mechanism based on direct signaling of this sort would have to incorporate the idea of the reaction forces traveling backward in time, and although such schemes have been tried (most notably by cosmologists Fred Hoyle and Jayant Narlikar), they have not been widely accepted.

Most supporters of Mach's principle today attempt to incorporate it into cosmology, not in terms of signal and response, but as part of the gravitational boundary conditions of the Universe—that is, as a statement about how the gravitational field of the Universe is organized as a whole. Einstein, who, as we have seen, was greatly enamored of Mach's principle, tried to formulate it in this sort of way as part of the general theory of relativity. After many decades of study, relativists have generally concluded that Mach's principle can be incorporated in the theory only by assuming that the Universe is spatially closed and finite. The simplest way to explain what is meant by this idea is by analogy with the surface of a sphere, such as the surface of the Earth. Our planet has a definite area, wrapped around a roughly spherical volume, but it has no edges; travel far enough in any direction on the surface of the Earth, and you get back to where you started. In that sense, the surface is "closed," yet unbounded. If the Universe as a whole were closed, one could envisage three-dimensional space to be "wrapped around" in some higher dimensional sense, so that, in this case, there would be a finite *volume,* with no edges. And it would be true, just as it is on the surface of the Earth, that if you traveled far enough in any direction you would end up back where you started.

But although it seems that Mach's principle can be made to work *only* in a closed universe, even closed universes do not necessarily incorporate the principle. In general, the theory of relativity is *not* consistent with Mach's principle, and in 1949 the mathematician Kurt Gödel, of the Institute for Advanced Study, Princeton, found a solution to Einstein's equations that describes a rotating universe. This does not mean that the actual Universe we inhabit is rotating, but it does show that Mach's principle is not "built in" to the equations of relativity, since according to Machian ideas the concept of the Universe rotating as a whole is meaningless—relative to what can the Universe as a whole be said to rotate? In this respect general relativity, despite its name, is actually closer to the spirit of Newtonian absolute space than to the relative motion of Berkeley and Mach.

And yet, the theory does predict several Machian-type effects. One such was spotted by Einstein himself, who wrote to Mach about it. Einstein reasoned that if rotation of an object were to be considered as relative to the Universe of material objects as a whole, then every object in the Universe would exert some influence on the rotating object. Most of the centrifugal effect would be due to the very distant matter in the Universe, but a small effect would arise from massive objects in the vicinity of the rotating object. Einstein then envisaged a particle of matter located inside a heavy spherical shell that is set spinning (relative to the distant stars) at great speed. To the extent that the shell contributes a tiny part of the Machian influence of the entire Universe, it should produce a small but significant force on the particle inside the shell, a force that has the effect of dragging the particle around in the direction the shell is rotating.

Remarkably, it is possible that similar effects may actually be measured soon. William Fairbank, of Stanford University, long ago proposed a space-borne gyroscope experiment that would orbit the Earth and measure the equivalent minute dragging effect caused by the rotation of our planet. According to Newtonian theory, such a gyroscope ought to point to a fixed position relative to the distant stars; but in Einstein's theory the rotation of the Earth leaves an imprint like a twist in its gravitational field, which reaches out into space and pulls the orbiting gyroscope around in the direction of rotation. An experiment to test this prediction may fly on the Space Shuttle in the 1990s. But even if the relativistic effect is seen, that will not on its own prove that Mach's principle is correct.

Mach's principle remains a compelling but elusive concept. Its fascination arises from the way it weaves the Universe into a unity, and relates otherwise insignificant pieces of matter to the great cosmic pattern. It is difficult to see how it could ever be verified by observations, but on the other hand it could be proved false if it were ever discovered that the Universe as a whole does possess an orchestrated rotation (relative, that is, to the frame of reference in which centrifugal forces vanish). This would show up in the cosmic microwave background radiation, left over from the big bang in which the Universe was born. This radiation, which permeates space, carries an imprint of all large-scale motions, and a cosmic rotation would appear in the form of a temperature variation of this radiation from different regions of the sky. In fact, observations show the microwave radiation to be remarkably uniform, and it is possible to place a very strong upper limit on the permitted rotation rate of the Universe. It turns out that *if* the Universe is rotating, its angular speed is so slow that it cannot have moved through more than a few degrees in its entire history.

For those who reject Mach's principle, these observations pose a mystery. There is no obvious alternative reason why the rotation rate of the Universe should be zero. Expressed another way, if rotation is absolute then it is purely an accident, a cosmological coincidence, that the frame of reference in which centrifugal force vanishes coincides, to very great accuracy, with the frame of reference defined by distant galaxies. This coincidence has, however, been addressed by cosmologists in recent years, along with several other coincidences of cosmology, through a development of the big bang theory of the birth of the Universe, known as inflationary cosmology.

Before we tackle the topic of inflationary cosmology within the framework of the new physics and our understanding of space and time, however, we must first take a look at the traditional picture of the Universe that emerges from the general theory of relativity. Lest this should appear too daunting a prospect to readers who feel that their understanding of relativity theory may not be quite up to the mark, we would first like to offer what we hope is a reassuring account of the way one of us came to grips with the key concepts. It happens that this is a firsthand account of the real-life experience of P.D., though both of us went through something similar.

Interlude: Confessions of a Relativist

There is an amusing story about Sir Arthur Eddington, who in the 1920s and 1930s was Britain's leading expert on Einstein's theory of relativity. Eddington was once asked to comment on the rumor that there were only three people in the world, by implication including himself and Einstein, who properly understood the theory. There was a long pause before Eddington replied, slowly, "I wonder who the third person is."

The fearsome reputation of the theory of relativity is often commented upon, and there is a widespread belief that any theory formulated by a man of such legendary genius as Albert Einstein must be beyond the power of ordinary people to grasp. Yet today, the theory of relativity is routinely taught in universities across the world, and libraries contain a range of student textbooks on the subject. Either the students of today are much brighter than they are sometimes given credit for, or the theory is not so fearsomely difficult to grasp, after all. But even so, it has to be said that many people do have the greatest difficulty in understanding these ideas, or in accepting that the world really does conform to some of the peculiar predictions of the theory.

My own struggle to master the theory of relativity began in 1960 when I was fourteen. The mathematician and science popularizer Sir Herman Bondi was invited to give a special lecture to pupils and parents at my school in London. The subject was "The Theory of Relativity." On the day of the lecture, Bondi's eloquent exposition proved wonderfully inspiring. Unfortunately, however, I got hopelessly lost on the technical details. Bondi's diagrams of space and time, showing all sorts of light signals going back and forth, left me utterly mystified.

Soon afterward, I discovered a book by Einstein himself, called *The Meaning of Relativity.* Alas, for all his genius as a mathematician, Einstein was a poor writer, and I found the book curiously unrewarding. The central idea did, however, sink in. This was that the speed of light is the same regardless of who measures it, or how they or the light source is moving. Such an obviously paradoxical result defied imagination, but being young and enamored of unusual concepts I accepted it uncritically.

Believing the impossible

As my education continued I came to learn of the various predictions of the special theory of relativity—the time dilation and length contraction effects, the impossibility of exceeding the speed of light, the rise in mass when a body is accelerated and the famous equation $E = mc^2$ expressing the equivalence of mass and energy. All these results I took to be true, but what they actually *meant* remained a puzzle to me.

At university I attended a proper course on the special theory. By that stage I couldn't avoid having to think through the time dilation effect in detail. It didn't just seem odd that someone could go off on a space trip and return to find their twin ten years older than they were; it seemed downright absurd. How could the *same* things happen at *different* rates? I asked myself. I formed the impression that speed somehow distorts clock rates, so that the time dilation was some sort of illusion—an *apparent* rather than a real effect. I kept wanting to ask which twin experienced "real" time and which was deluded.

It was at this point that I discovered the main obstacle to my progress. The trouble had been that I had kept trying to refer everything back to common sense and preconceived notions about reality, and this wouldn't work. At first, this seemed a shocking failure. I had to admit that I could not visualize time running at two different rates, and I took this to mean that I did not understand the theory. To be sure, I had learned how to manipulate the formulas and to calculate by how much differently moving clocks got out of step. I could work out *what* would actually happen, but I had no understanding of *why* this should be so.

It was then that I realized why I had been confused. So long as I could imagine the time dilation and other effects actually happening, and could work out the quantities involved, that was all that was needed. If I could always relate everything back to specific observers and ask what they would actually see and measure, then their observations would *be* the reality. This pragmatic approach of merely inquiring about what is observed and not trying to formulate a mental model of what *is,* in some absolute sense, is called positivism (see Chapter 1), and I have found it to be of the greatest help in grappling with much of modern physics.

Having got time dilation out of the way, the really hard thing for me to understand now was the concept of spacetime as a four-dimensional continuum. I had often read that time was the fourth dimension, but this bald statement meant absolutely nothing to me. Indeed, it sounded just plain wrong. My most elementary perceptions of the world told me that space is space and time is time. Qualitatively, they are so fundamentally different from one another that I could never see how time could be a fourth dimension of space. For a start, space is something I can see all stretched out around me, whereas I am only aware of one moment of time "at a time." Furthermore, I can move about in space, but not in time.

The problem with my understanding lay in taking the statement that time is the fourth dimension too literally. The theory does *not* assert that time is a fourth dimension *of space.* It recognizes that time is physically distinct from space, but also acknowledges that time and space are so intimately bound together in their properties that it makes sense to describe them jointly in four-dimensional language. The four-dimensional spacetime that results has some pretty odd features, already discussed. For instance, the four-dimensional distance between two events on the path of a pulse of light is zero, irrespective of how far apart in space they are.

When I first encountered this I was terribly confused. How could two different places be zero distance apart? Once I realized that time is not a dimension of space, this problem evaporated as well, because to measure the four-dimensional distance between two points that are separated in *both* space and time, it is necessary, as we have seen, to *subtract* the time difference from the space difference, in such as way that along a light path they cancel to give zero interval. Time is thus distinguished from space by its negative contribution to four-dimensional distance. If one were dealing with space alone, different points would indeed have to be nonzero distance apart.

Visualizing the invisible

So far, so good. The puzzles and paradoxes of the special theory of relativity pale, however, beside those that come with the general theory. I had picked up the oddments of the general theory of relativity while still at high school. I knew it was a theory of gravity, and that it treated the gravitational field in terms of curved space, whatever that was. I tried, without success, to envisage space curving. I could readily imagine the geometry of a block of rubber being distorted, because rubber consists of stuff, but space is just emptiness. How can "nothing" be curved? And precisely *where* does it curve? A block of rubber can curve "in" space, but space isn't "in" anything!

At this stage I formed the impression that the curvature of space manifested itself by causing the paths of the planets to curve around the Sun. The Earth moved in its elliptical orbit, I believed, not because it was pulled by a gravitational force from the Sun, but because the Sun curved the space in its vicinity, and the Earth merely followed the straightest possible path through this curved space. This seemed to make some sort of sense, because the Earth's orbit is curved, and I knew that even light rays are bent by the Sun. That must be it, I thought. Curved space simply means that bodies follow curved paths. Easy!

But now came a new puzzle. The Earth's orbit is a closed path. According to the mental image I had constructed, this implied that space was somehow folded around completely, engulfing the Solar System in a way that would shut it off from the rest of the Universe. Clearly that could not be right. The curvature of the Earth's orbit was obviously far too great for it to be due to curved space.

The mistake I had made was a subtle one. The curvature involved is *not* one of space, but of *spacetime.* The difference is crucial. Viewed in spacetime, the Earth's orbit is not a closed ellipse, but a shape like a coiled spring, known as a helix (remember Figure 17). The curvature involves both space and time, and whenever the time part enters into the picture, Einstein's theory requires one to multiply it by the speed of light. This is a very big number, and it has the effect of enormously stretching the helix. So although the orbit is tightly curved in space, it is a very shallow curve in spacetime. My original picture did have one correct feature, though. Curvature can indeed be viewed in terms of the paths of moving bodies, but their paths must be envisaged in spacetime, not just space.

At last I seemed to be making progress in understanding relativity. By far the greatest crop of difficulties came, however, when I began to learn cosmology. Einstein was famous for his concept of a "closed but unbounded" universe, and this concept defeated my best attempts at visualization. I had still not really got used to the idea that spacetime could be curved without being curved in anything. Now I was expected to believe that the whole of space might curve right around so that it somehow joined up again on the far side of the Universe. The pictures didn't really help. Showing the surface of a sphere and saying that this was a closed but unbounded surface in two dimensions was all very well, but going from two to three dimensions wasn't the simple extension that the writers who presented this analogy seemed to imagine. After all, a two-dimensional surface can be curved in three dimensions of space, but what could three dimensions curve into? I was stumped by the same old problem.

In the end, my taste for science fiction helped me over these difficulties. By reading fiction you get used to picturing yourself in the place of the characters, seeing an unfamiliar world through their eyes, sharing their experiences. Even when reading about the impossible, you can still *imagine* what it would be like for certain things to happen. After all, I had no trouble putting myself, mentally, in the place of H. G. Wells's time traveler, even though I knew the story made no sense from the point of view of physics. If time travel was conceivable, why not a closed universe?

I remembered my resolution not to try to envisage an absolute reality, not to struggle for some sort of God's-eye view of the whole Universe from outside. Instead I would take the humbler perspective of some poor space traveler laboriously exploring his closed universe. What would he experience? Well, he would be able to travel always in the same direction and yet come back to his starting point; that is one of the strange properties of Einstein's closed but unbounded universe. Though I still couldn't imagine *how* space could be arranged in that way, I could nevertheless imagine my space traveler having that experience. It made sense. There was nothing logically wrong with that happening. And if all his experiences fitted together consistently, however bizarre some of them might be, then that collection of experiences could be considered to constitute a sensible reality.

I applied the same philosophy to the notorious problem of the expanding Universe. Like everyone else, I couldn't fathom how the Universe could expand everywhere, because it seemed to me that it had nothing to expand into. But I could still imagine what it meant to observe the expanding Universe *from within*. I envisaged observers out there on distant galaxies, examining the heavens, and each one seeing the other galaxies moving away. Again, there was nothing logically wrong with that, even if I could not imagine *how* it was happening.

The trickiest problem of the lot concerned the notion of so-called horizons. I knew that the farther away a galaxy was located the faster it receded from us, and that there was a certain distance—known as our horizon—beyond which we can see no galaxies at all (this important feature of the Universe is discussed more fully in the following chapter). For a long while, after I first encountered the concept, I confused this limit with the frequently mentioned "edge of the Universe," and imagined that no galaxies could be seen inhabiting the space beyond the horizon because there were no galaxies there, only an unending void. Eventually I came to realize that there is no "edge" to the Universe at all; all references to such an edge were misleading nonsense.

But this confusion was dispelled only to be replaced by another. I had read somewhere that it was impossible to see galaxies beyond a certain distance because those galaxies are receding from us faster than light. I remember one day sitting in the cafeteria of my college discussing the subject with a fellow student. How can galaxies travel faster than light? I protested. "Ah!" replied my colleague. "The speed-of-light limit is a result of the special theory. In cosmology you must use the general theory." That was no help, though, as neither of us had mastered the general theory at that time.

In fact, we were talking at cross purposes. True, you must use the general theory, but that does not permit faster-than-light travel either. The cause of all the muddle was my inability to think of movement except in the old Aristotelian way. For me, if a galaxy was receding from us, it had to be moving *through* space. But this involved the erroneous concept of space as a sort of substance lying at rest, with material bodies passing through it like goldfish swimming in a bowl of water. The concept was simply wrong. It took a long time for me to hit on the picture that the expansion of the Universe is caused not by the assemblage of galaxies expanding out through space, but by space itself expanding, so that the gaps between the galaxies stretch.

I think I did not fully appreciate this idea of space stretching until I read about Willem de Sitter's model universe, which consists of nothing but expanding empty space. There is no matter in it at all! Of course, I had my usual difficulty in trying to imagine how space could expand, but viewed (as ever) in terms of what observers would actually see, it made perfect sense. Two observers would see each other swept apart by the expansion. Their mutual recession would be the reality. It didn't matter that I couldn't imagine how space, which didn't have any substance, could stretch in this way, so long as the observational consequences were consistent.

Armed with this new picture, the problem of faster-than-light travel did not occur. The galaxies were really not *traveling* at all, I realized. They were simply caught up in the general expansion of space. The famous red shift by which we know of the expansion was not, as I had read so many times (to my added confusion!), a simple Doppler effect of the sort that makes a train whistle drop in pitch when it passes by at speed. Instead, the light from distant galaxies is red-shifted because it comes to us across an expanding gulf of space, and the waves get stretched in transit. Eventually they get stretched so much that the waves can't be seen—the frequency is too low. This marks the horizon. The Universe beyond is still there, but it is invisible to us.

The dazzle of infinity

Perhaps the hardest thing of all for me to understand was the nature of the big bang in which the Universe was born. My initial image was of a very concentrated lump of matter sitting at a point in space. At some instant, for some reason, the lump exploded, sending fragments rushing away at great speed, eventually to become the receding galaxies. I now know that this concept was totally wrong, but in self-defense I would point out that my first brush with big bang theory came before the concept of spacetime singularities had been fully clarified, by Roger Penrose and Stephen Hawking, in the late 1960s.

At that time, the people studying this topic started to assert that the Universe had its origin in such a spacetime singularity, which was a point where spacetime became infinitely curved and where the laws of physics ceased to apply. It was not possible, so they said, for space and time, or any physical influence, to be continued through a singularity, so the problem of what existed before the big bang did not arise. There was no "before" because time began at the singularity. Nor, for the same reason, was it profitable, or even meaningful, to discuss what caused the big bang.

Later, I tried to picture a singularity by imagining all the matter in the Universe squashed into a single point. Of course, the very idea seemed outrageous, but I could imagine it. I was careful not to fall for the mistake of envisaging the point mass surrounded by space, however; I knew space would have to be shrunk to a point as well. This image worked well enough for a finite, closed model universe of the sort Einstin had invented, for we can all imagine something finite in size being shrunk to nothing. But there was an obvious problem if the Universe is spatially infinite. If the initial singularity was just a point, how could it suddenly change into infinite space?

I suppose infinity always dazzles us, and I have never been able to build up a good intuition about the concept. The problem is compounded here because there are actually two infinities competing with each other: there is the infinite volume of space, and there is the infinite shrinkage, or compression, represented by the big bang singularity. However much you shrink an infinite space, it is still infinite. On the other hand, any finite region within infinite space, however large, can be compressed to a single point at the big bang. There is no conflict between the two infinities so long as you specify clearly just what it is that you are talking about.

Well, I can say all this in words, and I know I can make mathematical sense of it, but I confess that to this day I cannot visualize it.

The subject that brought the general theory of relativity to world attention and really caught my own imagination was undoubtedly black holes. These bizarre objects have several weird properties that tax one's powers of visualization to the limit. When I first learned about black holes, in the late 1960s, I could accept the idea that a body such as a star might collapse under its own gravitation, and that this could trap light, causing the object to appear black. What I could not understand was what happened to the material of the star. Where did it go? Some theorems demonstrated that a singularity would form inside such a hole, but they did not demand that the infalling matter must encounter the singularity. If the matter misses the singularity, it cannot come back out of the black hole again, for nothing can escape from such an object. The situation thus seemed to me to lead to a contradiction.

The answer that I was given to this conundrum was that the

matter must go into "another universe." This sounded very exciting and profound. But what did it actually mean? Where was this other universe situated? I had mastered the idea of expanding spaces and closed spaces, but multiple spaces left my mind reeling. This was a really tough one. Again I fell back on my strategy of not trying to take a God's-eye view and imagine what these two spaces would look like side by side. I dealt only with what could in principle be observed from within these spaces.

I once read a short story called "The Green Door," in which a man comes across a door that leads to a very beautiful and peaceful garden that is something like our image of paradise. When he leaves, he cannot find the door again, and spends all his life searching for it. One day he sees a green door, goes through it, and falls to his death. The garden of the story did not exist in the space that we normally experience. The door was a link into another space. This must be, I concluded, the way to view the black hole. I could readily imagine the man's experience with the door, so why could it not be like that in the case of the black hole? You could go through the hole and come out somewhere that wasn't located anywhere in our space. I didn't need to know *where* this other space is, only that the experiences of an observer were logical and consistent.

Having related this little tale, however, I should caution the reader that, as we shall discuss in Chapter 9, the experts do not believe that you can actually go through a black hole like that. More probably all infalling matter *does* encounter the singularity, although this has not yet been proved to be the case.

Today, I have grown used to dealing with the weird and wonderful world of relativity. The ideas of space warps, distortions in time and multiple universes have become everyday tools in the strange trade of the theoretical physicist. Yet in truth I have come to terms with these ideas more through the familiarity of repeated use than through the acquisition of an esoteric power of intuition. I believe that the reality exposed by modern physics is fundamentally alien to the human mind, and defies all power of direct visualization. The mental images conjured up by words such as "curved space" and "singularity" are at best grossly inadequate models that merely serve to fix a topic in our minds, rather than informing us of how the physical world really is.

The situation resembles the world of international economics. We read about the US budget deficit of so many billions of dollars, and we think we know what that means, but in fact none of us can really envisage such huge sums of money in everyday terms. The words have a sort of pseudo-meaning, giving us something to latch on to while we pass on to the next point in the discussion, but not really conveying anything truly meaningful. It seems that if an idea is repeated often enough, then however counterintuitive it may be, people eventually come to accept it and to believe that they understand it.

The realization that not everything that is so in the world can be grasped by the human imagination is tremendously liberating. The theory of relativity still holds many technical mysteries for me—certain aspects of rotation and gravitational waves I find particularly tricky to understand. Yet having learned to overcome the need for simple images I can tackle such topics without fear. With mathematics as an unfailing guide, I am able to explore the territory beyond the boundaries of my own meager imagination to produce meaningful answers about things that can be observed.

Eddington's implicit boast of being the only person other than Einstein able to understand the general theory of relativity did not mean, I believe, that he and Einstein alone could visualize the revolutionary new concepts such as curved spacetime. But he may well have been among the first physicists to appreciate that in this subject true understanding comes only by *relinquishing* the need to visualize. Which may be a helpful thing to keep in mind, as we look at what relativistic cosmology has to tell us about the *observable* behavior of the Universe.

4 The Universe at Large

It is the job of the astronomer to study the various objects that make up the Universe. These include the Sun and planets, the many different types of stars, the galaxies and the interstellar material. By contrast, the cosmologist is less concerned with the specific cosmic furnishings, more with the overall architecture of the Universe. Cosmology deals with how the Universe as a whole came to exist, and how it will end. By "the Universe" cosmologists mean everything: the entire physical world of space, time and matter. Cosmology thus differs from other sciences in that its subject matter is unique—there is only one Universe to observe, and although cosmologists do sometimes refer to other universes (with a small "u"), they are talking about different mathematical possibilities which, like Gödel's rotating universe, may bear little relation to the real world.

Cosmologists draw upon the work of astronomers to build up a picture of the cosmos. They also use the laws of physics to model changes that occur as the Universe evolves and to try to predict its ultimate fate. And cosmologists today are increasingly willing to contemplate the initial conditions of the Universe, in addition to the laws themselves. Cosmology began as a serious science, though, only in the 1920s, with the discovery by Edwin Hubble that the Universe is expanding—the discovery that matched the predictions of the general theory of relativity, predictions that Einstein himself, believing the Universe to be static, had tried to squeeze out of his theory. The combination of Hubble's observations and Einstein's theory led to the profound conclusion that the Universe cannot always have existed, but must have appeared abruptly, several billion years ago, in the gigantic explosion we now call the big bang. Much of the effort of modern cosmological research is, as we have intimated, directed toward securing an understanding of the early stages of the Universe following the big bang, and in trying to relate observed features of the Universe today to physical processes that took place during that primeval phase.

Expanding without a center

Cosmology would not exist as a well-defined subject, though, were it not for the fact that we can talk of "the Universe" as a coherent reality. This in turn depends upon an important observational fact. On a large scale, matter and energy are spread remarkably evenly throughout space. "Large scale" here means on sizes greater than that of a cluster of galaxies, typically more than 100 million light-years; this uniformity implies that the Universe would look much the same from any other galaxy as it does from our own. There is nothing special or privileged about our location in the Universe. Moreover, the uniformity is maintained with time, so our galaxy shares a common cosmological experience with other galaxies as the epochs go by.

How does this relate to the notion of an expanding Universe? How, indeed, do we know that the Universe is expanding? The most direct evidence comes from examining the quality of the light received from distant galaxies. It was found by Hubble and others that this light is systematically shifted from the blue toward the red end of the optical spectrum. This means that the light waves are stretched somewhat compared with light from similar sources (the same kinds of atoms) in laboratories on Earth. Such a "red shift" is a familiar sign to physicists on Earth that the source of light is moving rapidly away from the observer, and this is how Hubble interpreted it. He concluded that the galaxies are rushing apart at great speed. As we have seen, this matched the fundamental requirement from the equations of general relativity that the Universe could not be static.

The galaxies are sometimes described as the fundamental building blocks of cosmology. It is their recessional motion that defines the expanding Universe. Within a galaxy, there is no expansion. Our own galaxy, the Milky Way (or just the Galaxy), consists of about 100 billion stars arranged in a flat disc, slowly orbiting the galactic core. The Milky Way is typical of a class of galaxies known as spirals, or disc galaxies, from their shape. Other forms are also known, but to a cosmologist variations between galaxies are minor details of the Universe. There is a tendency for galaxies to aggregate in clusters (anything from a few to a few thousand galaxies), attracted by their mutual gravitation, and this is of more interest to cosmologists. Because this tendency opposes the general expansion, it is really more accurate to envisage the *clusters,* rather than the individual galaxies, as the basic cosmological units.

Hubble spotted that the fainter galaxies visible to terrestrial telescopes had the larger red shifts. Because a galaxy appears fainter the farther away it is, he interpreted this to mean that more distant galaxies are receding faster. Later studies confirmed this, and showed that the velocity of recession, revealed by the red shift, is proportional to the distance of a galaxy from us. In other words, a galaxy twice as far away recedes at twice the speed, a relationship now known as Hubble's law. The number that determines just how fast a galaxy at a particular distance is receding is a key cosmological parameter, called Hubble's constant. Although its exact value cannot be determined from our limited observations of the Universe, most astronomers accept a figure of about 50 kilometers per second per megaparsec. A megaparsec is 3.26 million light-years, and this value of Hubble's constant means that a galaxy 10 megaparsecs away recedes from us at 500 kilometers per second, and so on.

In the beginning

This simple relationship between distance and velocity has a deep implication for the nature of the expanding Universe. It means that the Universe is expanding at the same rate everywhere: viewed from any other galaxy, the pattern of motion would be broadly the same. It is wrong to imagine, as many people do, that we are somehow at the center of the expansion. Although the other galaxies are certainly moving away from us, they are also moving away from each other, and because these motions obey the Hubble law the galaxies visible from any other galaxy recede from it in much the same way as the galaxies we can see recede from us. *No* galaxy is in the privileged position of being at the center of expansion.

If you have difficulty with this idea, it is helpful to consider the analogy of a rubber sheet, covered in dots to represent the galaxies. Imagine that the sheet is being stretched by pulling evenly in all directions around the edges (Figure 20). The effect is to cause every dot to move away from every other exactly as in the case of galaxies in the expanding Universe. Moreover, the system obeys Hubble's law: dots separated by twice the distance recede from each other at twice the speed.

Now it could be objected that the dots on the sheet *are* moving away from a common center, namely the middle of the sheet. If, though, the sheet were so large that you could not see the edges, there would be no way of knowing which dots were near the middle and which were not, merely by

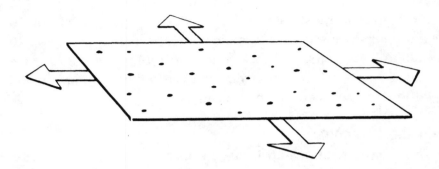

Figure 20. The expanding Universe can be likened to a rubber sheet being stretched uniformly in all directions. Here the sheet represents space (only two dimensions can be shown) and the dots represent galaxies. As "space" expands, so the distances between the galaxies grow, but the galaxies are not moving away from a common center, nor are they moving through "space."

inspecting their relative motions. If the sheet were *infinite* in extent, then there would truly be no meaning to the concepts of center or edge. In the real Universe there is not the slightest hint that the assemblage of galaxies has an edge anywhere, so there is no reason to talk about a center to the Universe or a region away from which the galaxies are rushing.

Nevertheless, one is still tempted to ask whether there *really is* an edge out there somewhere, beyond the range of present-day telescopes. After all, we cannot be certain that there are galaxies populating the Universe all the way to infinity. But even if the Universe is not infinite in extent, merely enormously huge, there is a sense in which speculation about a very distant edge to the Universe is pointless, if not meaningless. As the recession speed of galaxies grows with distance, there comes a point at which this speed is so great that it exceeds the speed of light. As the "confessions" of the previous chapter should have made clear, this does not break any of the relativistic rules—and the rubber sheet analogy also helps to make this clear. Although each dot moves when the sheet stretches, it does so *only* because the sheet stretches. There is no motion of dots *through* the rubber fabric. In the same way, it is better to think of the space between the galaxies as swelling or expanding, carrying the galaxies farther apart, rather than of the galaxies moving *through* space. This elasticity of space, a feature of general relativity, allows galaxies to effectively separate from one another faster than the speed of light, without any galaxy *passing by* another at this speed (which would violate the rules of special relativity). Thus, the red shift is produced because in the time it takes light to travel from a distant galaxy to Earth, the intervening space expands somewhat, and the light wave expands with it.[1]

Clearly we could not observe galaxies that recede faster than light, for their radiation could never reach us. So we cannot see beyond a certain distance, however powerful our telescopes. The limit in space beyond which we cannot see, even in principle, is called our horizon. As with the terrestrial horizon, its existence does not mean that nothing lies beyond, only that whatever might be over the horizon cannot be seen from where we are. There is certainly no edge to the Universe out as far as our horizon. Any more distant edge that might exist is in principle unobservable from Earth (at least in this epoch), so we might as well forget it. It is irrelevant to the *observable* Universe.

But it may well be that there is *in principle* no edge to the

1 So the cosmological red shift is *not*, in fact, produced in quite the same way as the Doppler shift we see in the light from moving objects on Earth, although in both cases the red shift is a signature of recessional motion. For nearby galaxies such as those studied by Hubble the two explanations for the red shift are, however, essentially equivalent.

Universe. The stretched rubber sheet that we have described so far is equivalent to flat spacetime—apart from the expansion, it is the same kind of flat space that the Greek geometers studied. If, though, the rubber sheet is curved into a spherical surface, like a balloon, then we can still imagine each dot on the balloon to represent a galaxy (or a cluster of galaxies) and we can still imagine the sheet stretching as the balloon expands, carrying every "galaxy" away from every other "galaxy" (Figure 21). There is no edge to such a model of the Universe, just as there is no edge to the Earth. This kind of model of the Universe is said to be "closed" for obvious reasons; equally obviously, the alternative of a universe that stretches away to infinity is called "open."

Is there any evidence to indicate whether the Universe is open or closed? In principle we can tell by carrying out some geometrical measurements. Geometry is different, remember, in curved space from the Euclidean geometry of flat space, and just as measuring the angles of large triangles drawn on the surface of the Earth could tell us that the Earth is roughly spherical, so measuring angles and volumes of shapes enclosing huge regions of space could in principle tell us how space is curved on the large scale. Such effects have been looked for (for example, by counting numbers of galaxies within spherical volumes of space of increasing radius), but other effects mask the one of interest.

There is, however, a more promising, though indirect, way of determining whether the Universe is open or closed. The presence of matter is what determines how space curves, and the more matter there is in the Universe, the more its gravity curves the space between the galaxies. There is a certain critical density, equivalent to about one atom of hydrogen in every liter of space (about 10^{-30} grams per cubic centimeter), which marks the boundary between space being

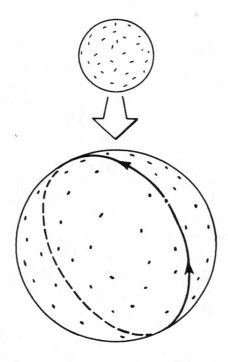

Figure 21. It is possible that space is closed into a finite volume without boundaries. In two dimensions, this is represented by a spherical elastic surface, like the skin of a balloon. The rubber of the balloon is "space" and the dots are "galaxies," as in Figure 20. The expansion of the Universe is analogous to blowing up the balloon. The line in the lower picture represents a path through space that completely "circumnavigates" the Universe.

open and closed. If the density is any greater than this, then, according to the theory of relativity in its usual form, space must be closed.

Observational evidence—by which we mean counting the number of galaxies in a given volume of space—suggests that the density of the Universe is substantially less than the critical value. But we also know, from the way that galaxies move within clusters, and the way that stars move within galaxies (both largely unaffected by the expansion of the Universe), that there is a great deal more matter in the Universe in some unknown form, tugging on the luminous galaxies gravitationally, even though it cannot be seen. As far as observations are concerned, we cannot say at this time whether the Universe is open or closed, but only that it sits very close to the dividing line. Investigations of the initial conditions of the Universe suggest, however, that the Universe *must* be closed, on theoretical grounds; as we shall see in Chapter 5, the inflationary model of the big bang also makes a strong prediction on this topic.

First, though, we should explain in more detail just what is meant by the cosmological notion of a big bang. It is obvious that if the galaxies are moving apart, they must have been closer together in the past. Extrapolating this trend, it seems that there must have been a time when all the matter in the Universe was compressed together. Among the many popular misconceptions about the nature of the big bang and the expanding Universe is the notion that this original lump of matter was located somewhere in a preexisting void, and that fragments of this "primeval egg," blown apart in the big bang, are now flying apart from a common center into the space that surrounds it. But as we have explained, the expansion is better envisaged as that of space itself, carrying the galaxies along for the ride. So when all the matter of the Universe was gathered together, that was because the space between the galaxies was shrunk (or rather, not yet expanded). Space itself, and time, were created, like matter, in the big bang; there was no "outside" into which the explosion occurred.

From the Hubble law, we can deduce the rate at which the Universe is expanding, and calculate backward to find the time when space began to expand, the time when all matter was squeezed together in one place. The simple Hubble law tells us that this was many billions of years ago; there is, however, a subtlety that must be taken into account. The Universe does not expand freely, but is subject to gravitation. The presence of all the matter in the Universe continually slows the rate at which it is stretching. So the Universe must have been expanding more rapidly in the past than the rate at which we see it expanding today. Taking this into account, the time when the expansion began—the big bang— works out at about 15 billion years ago.

This slowing down of the universal expansion also has another interesting effect. As the expansion rate slows, galaxies that used to be moving "faster than light," from our point of view, slow to a speed less than that of light. The effect of this is that the universal horizon grows outward as time passes; the horizon itself (moving, in a sense, at the speed of light) expands *faster* than the Universe, so that as time goes by it encompasses more and more galaxies, even though the galaxies continue to retreat from us.

If this picture of the expanding Universe is taken literally, and "wound back" far enough, then the volume of space encompassed by the position of our horizon today, and all other finite volumes of space, were shrunk to zero "in the beginning." This implies that the Universe was infinitely compressed, with all the cosmic material we can see squeezed together into a single point. Cosmologists use the term "singularity" to denote this limiting state. According to the general theory of relativity, such a singularity represents a boundary to space and time. It is not possible to continue space and time back through the singularity, which, in that sense, *does* represent the edge of the Universe, though it is an edge in time rather than in space. For this reason, the big bang is taken to represent the origin of the entire physical Universe, and not merely the origin of matter.

The question "What happened before the big bang?" becomes meaningless, since there was no "before." And the similar question "Where did the big bang occur?" can be answered: everywhere! There is no center or edge to the Universe, in the everyday sense; the explosion did not occur *in* space, at some particular location. The big bang was the explosive appearance *of* space.

This is such an important point, and so often misconceived, that we wish to make it even more clear, with a return to the balloon analogy. Imagine the radius of the balloon being progressively shrunk; this corresponds to going backward in time toward the big bang. The fabric of the balloon represents space itself, and as the balloon gets smaller, so there is less and less space. In the limit that the balloon shrinks to zero radius, the area of the balloon's surface dwindles to nothing, and the "Universe," space and all, simply disappears at a point. Returning to the "forward time" description, the big bang was the abrupt creation of the Universe from literally nothing: no space, no time, no matter.

Time and the Universe

This is a quite extraordinary conclusion to arrive at—a picture of the entire physical Universe simply popping into existence from nothing. The conclusion is based on an idealized picture, in which the Hubble law is taken as applying precisely to a model of the Universe in which matter is spread with precise uniformity. In fact, the Universe is not precisely uniform; matter is aggregated into, for example, galaxies. Moreover, it seems likely that the expansion rate is not *exactly* the same everywhere and in all directions. At first sight, it might seem that these imperfections may invalidate the conclusion that there was a singularity marking a past limit to the Universe, for if one were to follow back in time the careers of different portions of the Universe, then, perhaps, they would not come together in *exactly* the same place at the same time—the singularity might, we may guess, be smeared out, in some sense. Remarkably, though, it can be shown that even in an irregular Universe a singularity of some sort is inevitable—so long as the force of gravity is always attractive.

That solitary loophole has encouraged some cosmologists to conjecture that perhaps, under the extreme conditions of the big bang, something like antigravity is possible. This could remove the singularity. A possible scenario would then develop along the following lines. Before the big bang, the Universe was contracting, falling together under its own gravity. At some stage when the density was very high, gravity turned into antigravity, causing the Universe to "bounce," and initiating the present phase of expansion.

But this removes one problem at the expense of introducing another. If the Universe did not come into existence a finite time ago at a singularity, then it must always have existed. This means that physical processes have been going on for all eternity. But essentially all physical processes that we observe in the Universe are finite and nonrenewable. Stars, for example, do not shine forever. Eventually, they burn up all their fuel and collapse, perhaps into a black hole. The supply of material for new stars is limited, so these irreversible processes cannot have been going on forever.

It might be countered that the high-density "bounce" phase would pulverize and reprocess matter, destroying all traces of earlier systems and structures, and thereby renew the Universe. But this runs into conflict with a fundamental law of physics known as the second law of thermodynamics, which places strong constraints on what can be achieved by *any* cyclic physical process. In particular, as we shall shortly explain, it forbids any process from returning the Universe as a whole to an earlier physical state. For these reasons, most cosmologists prefer to believe that the Universe has a finite age, and that the big bang really did mark a creation from nothing. A corollary to this is that if the Universe had a definite beginning ("birth"), then it must surely have a definite end ("death").

Is the Universe dying?

The answer to this question is intimately bound up with the science of thermodynamics, and with our understanding of the nature of time—for, clearly, however much different observers may disagree about the mysterious moment of "now," if the Universe was born in the big bang and will die at some future date, then we have a fundamental indication of the flow of time from past (birth) to future (death).

The notion that the Universe might be dying originated with the German physicist Hermann von Helmholtz, in 1854. He announced that the Universe was indeed doomed on the basis of the then nascent science of thermodynamics, in particular the second law, which proclaims the ultimate triumph of chaos over order. Helmholtz envisaged a universe which began in a relatively well-ordered state, and then slid slowly and inexorably toward what was called a "heat death," a situation of thermodynamic equilibrium in which all sources of useful energy are exhausted and nothing of further interest can occur. This one-way slide from order into chaos imprints upon the physical world an arrow of time sharply distinguishing past from future, an arrow that is apparent in everyday experience from the fact that things wear out—cars rust, people grow old, and so on. This arrow points in the same direction as the cosmological arrow of time, away from the big bang, although, of course, Helmholtz knew nothing of this model of the birth of the Universe when he formulated his ideas.

A simple example of irreversible change from order into chaos occurs whenever a new deck of cards is shuffled. Starting with the cards in numerical and suit order, it is easy to jumble them up. But days of continuous shuffling would fail to reorder them, even into separate suits, let alone numerically. We could make a movie of someone shuffling such a deck of cards displaying the sequence before and after the shuffle; running the movie backward would not fool anybody, if the cards were initially in order. This is a good operational definition of time asymmetry—if a reversed movie looks impossible, the arrow of time is at work. If the deck was already well shuffled before we began making the movie, however, then it would look just as plausible either way; once chaos has been achieved, there is no further significant change and, in a sense, time ceases to flow.

It is possible to give a precise quantification of the degree of disorder in a physical system. This is called *entropy*. In a closed system, entropy never decreases. The qualification "closed system" is vital. In open systems, entropy can decrease, but the increase in order in the open system is always paid for by a decrease in order (increase in entropy) somewhere else. In the growth of a crystal, for example, the ordered deposition of ions in a lattice produces heat which flows away into the environment, raising its entropy. There is thus no incompatibility between the inference that the Universe as a whole is slowly dying as its overall entropy rises, and the manifest growth of order (decrease in entropy) in certain systems, such as growing crystals or biological organisms.

The first systematic investigation of the arrow of time was

made in the 1880s by Ludwig Boltzmann, who studied the statistical behavior of large numbers of molecules. His equations seem at first sight to prove that the entropy of a box of gas would always go up if the molecules of the gas cavorted at random. That is, random encounters between molecules always "shuffle" the gas into a more disorderly state. But this immediately suggests a paradox. The laws of motion which Boltzmann assumed for the gas molecules (Newton's laws) are symmetric in time. Every motion of the molecules (think of a collision between two pool balls on a perfectly smooth table) could, in principle, be reversed without violating these laws. Yet the time-reversed motion would lead, for a box of gas, to increasing order and decreasing entropy; how had Boltzmann smuggled time asymmetry into the collective activity of the molecules?

In fact, a box of gas molecules obeying Newton's laws *perfectly* does not have an inbuilt arrow of time. It is possible to prove that, over unimaginably long time intervals (far longer than the time since the big bang), the continual random molecular "shuffling" of the gas will cause it to visit and revisit every available state, much as continual random shuffling of a deck of cards would *eventually* reproduce any sequence, including exact numerical and suit order. What Boltzmann's calculations actually show is that if a gas is in an ordered state of low entropy at one particular moment, it will *very probably* soon be in a less ordered state, and will rapidly move toward an apparent state of equilibrium at which the entropy is at maximum. But this will not be *absolute* equilibrium. Statistical fluctuations will occur, and eventually the gas will find its way back to its original ordered state (with, in a sense, time running backward as the system becomes more ordered), after which another cycle will begin. Thus, the entropy of the gas will go down as often as it will go up, if we watch it for enormous spans of time.

What, then, is the origin of the arrow of time in the real world? The answer lies not in the laws of molecular motion but in the initial condition of the gas. Boltzmann proved that *if* a gas is in a relatively ordered state *then* its entropy should (*very* probably) rise; but the real issue is how it achieved the ordered state in the first place. In practice this is never a result of our waiting for an incredibly rare fluctuation to lower the entropy of the gas; it is because the Universe as a whole is progressing from a state of low entropy to a state of high entropy, and we are able to dump entropy into the flow going past. This makes it possible to set up a situation in which, for example, gas is confined within one half of a glass box, with a movable partition sealing it off from the other, empty half of the box (Figure 22). There is order in this system that is not present when the partition is removed and the whole box is uniformly filled with gas. The person who creates such a low-entropy state does so, however, at the cost of activities, such as manufacturing the glass box, which raise the entropy of the Universe at large. And the strictly local decrease in entropy is also strictly temporary. When the partition is removed (or the box breaks), gas rushes across to fill the whole box (or out into the world at large), destroying the order that was created and raising the gas's own entropy in the process.

All of this is possible because the Earth is an open system, through which energy and entropy flow. The source of almost all the useful energy we use is the Sun, which is a classic example of a system in thermodynamic disequilibrium—a compact ball of hot gas in a relatively low entropy state is irreversibly pouring huge amounts of energy out into the cold vastness of space. Encounters with time's arrow that we ex-

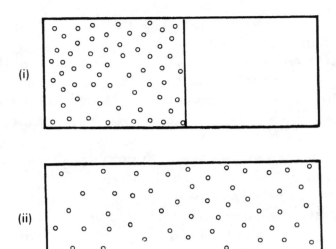

Figure 22. (i) A gas is confined in one half of a box. (ii) When the partition is removed, the gas expands to fill the whole box. State (i) is relatively more ordered than state (ii), so it has a lower entropy. The seemingly irreversible transition from the lower to the higher entropy state defines a thermodynamic arrow of time.

perience in everyday life have their origin in our proximity to this great source of energy in the sky, which is like a bucket of negative entropy into which we can dip in order to create ordered systems on Earth.

In order to pin down the origin of time's arrow, therefore, we have to discover how the Sun achieved the state of less-than-maximum entropy which not only allows but requires it to pour out energy in this way. And as the Sun is just one star among many, the problem is essentially a cosmological one: why is it that the Universe is in a state of disequilibrium, with a lot of heat energy concentrated in the stars, and not spread evenly through space?

The question is not new. It was asked, in a slightly different form, in the eighteenth century by the Swiss astronomer Jean-Philippe de Cheseaux, and again a century later by the German Heinrich Olbers, before finally being resolved in the twentieth century. The puzzle which baffled de Cheseaux and Olbers, among others, was that if the stars had been radiating heat and light forever, then the space between the stars should have been filled with radiation energy, so that the temperature of radiation in space should be comparable with that of the stars. If that were so, the night sky would blaze with light. Although it was not expressed in this way until modern times, there is something of a paradox in that space is so much cooler than the stars. Why has the Universe not come into thermodynamic equilibrium?

The answer to the question comes from looking, not at the laws of physics as they apply in the Universe today, but at the cosmic initial conditions. There were no initial conditions in the original formulation of the puzzle, precisely because it was assumed that the Universe was infinitely old; but we no longer share this view, and one of the most persuasive pieces of evidence that there *were* initial conditions is, indeed, the darkness of the night sky. The stars generate their energy by burning nuclear fuel, converting light elements (initially hydrogen) into heavier elements (starting with helium). If other events do not overtake it, a large star will continue to derive energy this way until its material is composed of the element iron, which is the most stable (that is, highest-entropy) nuclear material. In converting hydrogen to iron the star will have produced a huge increase in entropy, releasing energy—originally locked up in its nuclei—as radiation that spreads across the far reaches of the cosmos.

So we must look further back, to the origin of the hydrogen fuel that makes this possible. This takes us back 15 billion years or so to the big bang. Cosmologists infer, from the measured expansion rate of the Universe and the temperature of the cosmic background radiation today, that one second after the initial singularity the temperature of the Universe was 10 billion degrees. This was too hot for complex nuclei to exist, and the cosmological material consisted of a soup containing only the most basic atomic building blocks (individual protons, neutrons and electrons) and other "elementary particles." As the temperature fell, the atomic particles began to combine into composite nuclei, but only 25 percent formed helium, with less than 1 percent forming slightly heavier elements and almost 75 percent remaining in the form of simple hydrogen, by the time the process came to an end. This period of nuclear fusion lasted just a few minutes, and it terminated because after that time the temperature was too low for nuclear reactions to continue. As a result, most of the atomic stuff of the Universe was "frozen" in the form of hydrogen, and is still in that state—a low-entropy state—today. Only inside stars, squeezed to enormous pressures by the inward tug of gravity, where temperatures of several million degrees recreate in miniature the conditions a few minutes after the moment of singularity, can the nuclear reactions switch back on, and continue the inexorable slide of the Universe toward heat death. It is this residue of hydrogen fuel that ultimately drives most of the interesting activity in the Universe, the activity that imprints on the cosmos an arrow of time.

But now we are confronted by another puzzle. If the Universe began in a low-entropy state, from which it is irreversibly degenerating, then we would expect the early Universe to have been rather far from thermodynamic equilibrium (the latter being the state of maximum entropy). However, there is good evidence that the Universe at one second was in a state rather close to thermodynamic equilibrium. The background radiation itself, the smoothness of the distribution of matter on the large scale and the simplest interpretation of the relativistic equations all tell us the same thing. How did it come about that the Universe has apparently gone from equilibrium to disequilibrium when the second law requires the opposite trend? Or to put the problem more graphically, if we think of the Universe as like a huge clock slowly running down toward the inactivity of the heat death, what wound the clock up in the first place?

The answer lies in the expansion of the Universe. It was this expansion that caused the cosmic material to cool. A star like the Sun would *not* have stood out as an exceptionally hot object against the "background" temperature of the Universe a few minutes after the singularity. It stands out today not because the *Sun* is hot but because the *Universe,* thanks to that expansion, is cold. It is expansion that keeps space cool while the stars become hot. In this respect, the Universe is not a perfect example of a closed system, because it is continually being expanded, rather as if we kept moving the partition in our box of gas so that it could never settle into equilibrium. The expansion produces the essential thermodynamic disequilibrium that gives the arrow of time it direction.

But this answer is only satisfactory up to a point. The thermodynamic arrow of time is only one of several known, although they may be related. Another arrow concerns gravity. Gravitating systems have a natural tendency to progress from regular into irregular forms, such as when a smooth cloud of gas clumps into stars (Figure 23). The ultimate triumph of this one-way process is the black hole, where matter clumps so tightly that it irreversibly collapses out of existence. The fact that things can fall into black holes, but cannot come out of them, is a clear example of time asymmetry—the "movie"

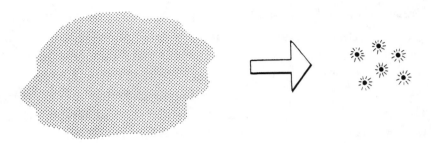

Figure 23. An initially smooth, extended cloud of gas will evolve, under the influence of its own gravity, to a highly inhomogeneous state in which the material is clumped into stars. This defines a gravitational arrow of time.

cannot be reversed (Figure 24). As the Universe continues its march toward the heat death, more and more matter will end up in black holes. Roger Penrose of the University of Oxford has pointed out that the entropy of the observable Universe is a mere 10^{-30} of the value it would have had if all the material in it were concentrated into a black hole. This raises the question: why did the big bang not simply cough out black holes? More generally, why was the early Universe in such a gravitationally regular state, consisting of uniformly expanding, almost perfectly smooth gas, when the most probable (highest-entropy) state consists of the irregular clumping of material in black holes? The evidence for this smooth early state of the Universe comes, as we have hinted, from the featureless state of the cosmic microwave background radiation. This radiation would carry an imprint of any irregularities present in the early Universe, but in fact, as we discuss in Chapter 5, it is uniform to within one part in 10,000.

To summarize the story so far, there seem to be at least three different arrows of time in the Universe: thermodynamic, cosmological and gravitational. Almost certainly, they are linked in some way. The low-entropy thermodynamic

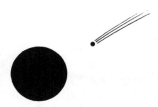

Figure 24. A black hole is the ultimate state of gravitational clumping. When an object falls into a black hole it can never escape. This is the most striking example of the irreversible nature of the gravitational arrow of time.

state of the Universe can be traced to the cosmic expansion; the cosmic expansion itself is an example of the gravitational activity of the Universe; and the general tendency for gravitating systems to evolve from smooth to clumpy is exemplified in the way that the cosmic expansion is so uniform and regular. So an explanation for the arrow of time seems to boil down to accounting for why the Universe started out in such a smooth and regular state. Is this simply because the Universe just happened to be "made that way"—is it, in other words, an arbitrary initial condition beyond the scope of science? Or is it possible to explain the initial smoothness of the Universe using a physical theory of the cosmic origin? Either way, we have traced the arrow of time back to the creation event itself, and the processes that took place in the split second that followed.

Before we leave our discussion of the arrow of time to tackle the topic of the very early Universe, however, we should say something about a rather different sort of paradox concerning the nature of time. Whatever the resolution of the puzzle about the origins of time's arrow, there is no denying that there is an arrow, and that this provides a distinction between past and future. But we have claimed that the theory of relativity has no place for past, present and future! How can these facts be reconciled?

Time and consciousness

As we explained when discussing the nature of simultaneity (Figure 14), the concept of a unified spacetime "continuum" implies that time is "stretched out" in its entirety, like space. No absolute and universal meaning can be attached to the notion of "the" present. Furthermore, the idea that time "flows," or that the present moment somehow moves from the past to the future in time, has no place in the physicist's description of the world. This state of affairs was neatly summed up by the German physicist Hermann Weyl, who declared that "The world does not *happen*, it simply *is."*

Many people confuse the existence of an arrow of time with the psychological impression that time is flowing or moving in one direction. This is due in part to the ambiguous symbolism attached to the idea of an arrow, which can be used to denote either motion in the direction of the arrow or simply a spatial asymmetry, as when the arrow on a compass needle distinguishes north from south. When a compass needle *points* north it does not mean that you are *moving* north. But confusion also arises as a result of a linguistic muddle over the use of the terms "past" and "future." The concepts of past and future do have a place in physics, provided that one is careful to use these words in the correct way grammatically. The notion of *the* past or *the* future is not allowed. Nevertheless, one can still talk about one event being in the past of *another event*. There is no doubt that events are *ordered* in time, just as the pages of this book are ordered in space, in a definite sequence; furthermore, this order, like the numbers on the pages of the book (and, we hope, the "flow" of our narrative), has a direction associated

with it, even if nothing actually "flows" at all. After all, the
very idea of causality demands that some sort of earlier-later
relationship applies to events. To take a simple example, if
you fire a gun at a target and the target shatters, there can be
no ambiguity about the order of events, to any observer; the
target shatters *after* the gun fires. The effect lies in the future
of the cause.

But when we refer to an "arrow" of time, we should not
think of the arrow flying through the void from past to future; rather, we should think of the arrow as like the compass needle, pointing the way to the future, even though it is
not *moving* into the future.

Philosophers have long debated the thorny issue of
whether the present moment is objectively real or just a psychological invention. Those, such as Hans Reichenbach and
G. J. Whitrow, who have argued for the reality of the present
are known as "A-theorists," while their opponents, among
whom are some distinguished figures such as A. J. Ayer,
J.J.C. Smart and Adolf Grünbaum, are called "B-theorists."
The terminology A and B reflects the existence of two quite
distinct modes of speech. The first, the so-called A-series,
uses the concepts of past, present and future, and the rich
vocabulary of tenses that permeates human language.[2] The
second system of discussing temporal sequences, the B-series, uses a system of dates. Events are labeled by the date on
which they happen: Columbus sets sail, 1492; first man on
the Moon, 1969; and so on. This serves to place events in
order unambiguously, and is the system that physicists use.

2 With, perhaps, one exception. We are told by experts in linguistics that the
Hopi people of North America do not have linguistic distinctions between
past, present and future, and have no way of expressing the idea of a flow of
time. For them, events are categorized only by whether they are "manifested"
or are "evolving."

The dates are simply coordinates, exactly analogous to the use of latitude and longitude for defining spatial positions on the surface of the Earth, and as far as the physicist is concerned that is all that is needed to give a complete account of the world.

B-theorists argue that these two schemes of discussing the *same* set of events cannot be compatible. Because the present moment is always moving ahead in time, events that start out being in the future, sooner or later become the present, and then the past. But a single event cannot be labeled by all three descriptions—it cannot be in the past, the present and the future.

A further difficulty, they point out, concerns the question of how *fast* the present moves forward in time—how fast time flows. The answer can only be one second per second (or twenty-four hours per day), which tells us nothing at all; it is a mere tautology. The concept of flux or change refers to something that has different values at different times. But what sense can one attach to the notion of *time* changing with time?

This problem was taken up some years ago by an imaginative writer named J. W. Dunne, who invented something he called serial time. Dunne accepted the idea that the present moves, but realized that this only makes sense if one introduces a second measure of time against which the flow of the first time can be gauged. By extension he then proposed a third time, a fourth time, and so on in an unending sequence. Dunne tried to link these different layers of time to the layers of our consciousness, and went on to suggest that during dreaming one's consciousness could move about in "Time 1," thereby witnessing both past and future events. Hardly surprisingly, Dunne's ideas have not been taken seriously by either scientists or philosophers; but they do illustrate the difficulties inherent in the concept of taking the notion of the flow of time seriously.

At this point, the skeptical reader may well protest. A typical argument goes as follows. "Whatever the physicists and philosophers may say, of course things *happen*. There *is* change; I experience it directly. For example, today I smashed a coffee mug: this event occurred at four o'clock, and it is a change for the worse. My coffee mug is now broken, and it wasn't this morning."

The B-theorist, however, will retort that there is only an illusion of change. "All you are really saying is that before four o'clock the coffee mug is intact, after four it is broken, and at four it is in a transitional state." This neutral mode of description—the physicist's B-series—conveys precisely the same information about the coffee mug events, but makes no reference to the passage of time. "There is no need to talk of the coffee mug *changing into* a broken state, or to say that this *happened* at four o'clock. There are simply dates, and states of the coffee mug; that is all. No more need be said."

Indeed, the B-theorist can go beyond this, by pointing out that we never measure time directly. What we actually measure is something physical, like the position of the hand on a clock or the position of the Earth in its orbit around the Sun. When we say that something was broken at four o'clock, what we are really saying is that intact states correlate with the little hand of the clock being above the number 4, and broken states correlate with the little hand being below the number 4. In this way, it is actually possible to eliminate all reference to time in describing the world.

The A-theorist might counter that the notion of the changing position of the clock hand itself requires a reference to time, unless it too is correlated with something, such as the rotation of the Earth. But then one can wonder about the motion of the Earth, and so on. What lies at the end of this regress?

Once again, we are forced to contemplate initial conditions. The ultimate clock is the Universe itself, which through its progressive expansion in size defines a "cosmic time." It seems as if there may be some deep significance in this—both the thermodynamic arrow of time and the philosophers' arrow of time seem to have their roots in the expansion of the Universe, in the cosmological arrow of time. But when we try to study the origins of this expansion of the Universe in terms of the best scientific description of mechanics, quantum mechanics, we are presented with a major surprise. Cosmic time drops out of the equations altogether! The gravitational equations that govern the motion of the cosmos impose a restriction (known technically as a constraint) which has the effect of eliminating the time coordinate. As a result *all* change must be gauged by correlations. Ultimately, everything must be correlated with the size of the Universe. Any vestige of a moving present has faded completely, exactly as the B-theorists have always claimed.

But what about the fact that we *feel* time flowing? Remember that Einstein spoke of an illusion. Illusions of motion are known in other contexts. A familiar one is giddiness. If you spin around rapidly and then stop suddenly, you have an overwhelming sense that the Universe is rotating. Yet logic and your eyes tell you that you are not moving. Might it be that our strong sense of the flow of time is an illusion akin to giddiness, perhaps connected with the way our memories operate?

The argument is far from being satisfactory. Although the great weight of scientific and philosophical argument is on the side of the B-theorists, and against the objective reality of a moving present, it seems impossible to shrug the matter aside. Surely there must be an aspect of time that we do not yet understand, which surfaces in a muddled and incomplete way in our perception of a moving present moment? We have already mentioned chaos, which removes the spirit of Newtonian determinism from our world view. In the sense that the future is unpredictable, it is not already fixed by the present. Another strand of research, which we describe more fully in another part of this book (Chapter 7), involves quantum theory, which tells us that there is an *inherent* uncertainty in the outcome of events at the subatomic level. In quantum mechanics, many possible future patterns of events all exist, in some sense, until observation converts the potential into the actual. This crucial transformation might just in some way be connected with the woolly concept of the flow of time.

Unsatisfactory though it may be, we have to admit defeat in our attempt to decide what time *is,* and to make do with our everyday images of the flow of time in trying to describe the origin and ultimate fate of the Universe. This very admission of defeat is, however, in itself one more indication of the need for a post-Newtonian paradigm, a sign that there is more to the Universe than our established scientific theories can yet encompass. Meanwhile, how far can twentieth-century science go in explaining the origin of space and time?

5 The First One Second . . .

In 1976, American physicist Steven Weinberg produced a book called *The First Three Minutes*. It described the early stages of the Universe, the big bang itself. But the title of the book was a little bit of a cheat. The story Weinberg told, about how a superdense state of primordial stuff became an expanding Universe in which atomic matter was distributed evenly across space in the form of roughly 25 percent helium and 75 percent hydrogen, did indeed end about three minutes after the singularity—but it also *began* a hundredth of a second after the singularity, not "in the beginning," itself. At that time, this was as far back as physicists could push their theories of the big bang, and what happened during the first hundredth of a second remained largely a mystery. Now, less than twenty years later, some theorists speak confidently of events that occurred during this first hundredth of a second. They still cannot push back all the way to the singularity, but not through any deficiency in their theories; it is now widely accepted that there is a fundamental unit of time, the "Planck time," beyond which intervals cannot be subdivided. This quantum property of spacetime implies that time "began," in a sense, when the Universe was 10^{-43} of a second "old." The singularity itself can never be probed. What had previously been treated as a singularity at the origin of time is smeared out by quantum effects. Our understanding of the first one second of cosmic history is now on a par with the understanding of the first three minutes achieved by cosmologists in the mid-1970s; and within that first one second there occurred the processes that smoothed out the observable Universe and set it running in a state of low enough entropy for interesting things, including ourselves, to appear at later dates.

The big bang implies the appearance, not merely of matter and energy, but of space and time as well. The bonds of gravity marry spacetime to matter; where one goes the other must follow. The big bang is the past extremity of the entire physical Universe, and marks the beginning of time; there was no "before." This perplexing concept was long ago anticipated by St Augustine, who maintained that the world was created "with time, not in time."

Generations of philosophers and theologians have argued about the meaningfulness of a creation "with time." Such an event must be without prior cause, for causation is itself a temporal concept. This cosmic conundrum sparked off endless and inconclusive debate about God's temporality. But modern physics, specifically the quantum theory, has thrown new light on the relation between cause and effect, cutting across the old paradox of what caused a big bang for which there is no "before."

For our present purposes, the central feature of the quantum theory is indeterminism. The old physics linked all events in a tight chain-mesh of cause and effect. But on the atomic scale the linkage turns out to be loose and imprecise. Events occur without well-defined causes. Matter and motion become fuzzy and indistinct. Particles do not follow well-defined paths, and forces do not produce dependable actions. The precision clockwork of classical Newtonian mechanics gives way to a ghostly melee of half-forms.[1] It is out of this

---

1 Although this is not the place to provide a detailed history of the development of quantum physics, we should, perhaps, stress that all these ideas, like the noncommon-sensical implications of relativity theory, have been amply

submicroscopic ferment that the essential quantum uncertainty emerges. What happens from moment to moment cannot be predicted with definiteness—only the betting odds can be given. Spontaneous *random* fluctuations in the structure of matter, and even of spacetime, inevitably occur.

Something for nothing

One of the more bizarre consequences of quantum uncertainty is that matter can appear out of nowhere. In classical physics, energy is a conserved quantity; that is, it can be neither created nor destroyed, only changed from one form into another. Quantum mechanics permits energy to appear spontaneously from nothing as long as it disappears again rapidly. Since matter is a form of energy, this provides, as we mentioned in Chapter 1, for the possibility of particles appearing briefly out of nothing. Such phenomena lead to a profound modification of what we mean by "empty" space.

Imagine a box from which all particles of matter have been removed. We might think of this as a perfect vacuum—empty space. In fact, the fluctuating quantum energy of the vacuum causes the temporary creation of all manner of "virtual" particles—particles that exist only fleetingly before fading away again. The apparently inert vacuum is actually a sea of restless activity, full of ghostly particles which appear, interact and vanish. And this applies whether or not the box is emptied of all "permanent" matter—the same restless vacuum activity goes on all around us, including in the space

confirmed as a good description of the way our Universe works by repeated experiments. Indeed, it was the failure of Newtonian physics to explain the observed results of certain experiments that pointed up the need for a new theory. Quantum theory *does* provide an accurate description of how things work on the subatomic scale.

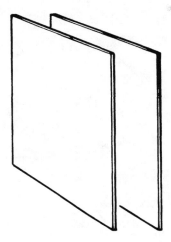

Figure 25. The Casimir effect. The parallel reflecting plates disturb the structure of the quantum vacuum in the space between them, by forcing the virtual photons to adopt only a limited set of wavelengths. The result is to produce a force of attraction between the plates.

between atoms of ordinary matter. Moreover, this irreducible vacuum activity is no mere theorists' speculation. It produces real physical effects on atoms and subatomic particles, effects that have been detected in many experiments. The Dutch physicist Hendrik Casimir proposed an experiment in which two metal plates are placed face to face, a short distance apart (Figure 25). Because they are metal, such plates will be highly reflective to photons, including the hypothesized virtual photons of the quantum vacuum. The effect of these reflections, bouncing back and forth between the two plates, changes the nature of the vacuum in the gap between the plates in a measurable way.

The simplest way to picture what is going on is by analogy with the vibrations of a plucked guitar string. Because the string is fixed at each end, it can vibrate only with certain

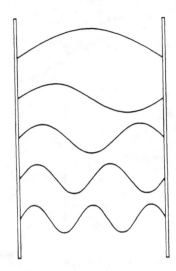

Figure 26. The virtual photons trapped between the plates of Figure 26 behave in a similar way to vibrating guitar strings. The lowest frequency of the string corresponds to one half-wavelength just fitting its length. The next allowed frequency (harmonic) has exactly one wavelength fitting the string, the next has one and a half wavelengths, and so on.

notes, as any musician knows. Vibrations that travel along the string reflect back from the fixed ends, so that the guitar string will play only a given musical note—the one for which exactly one half of a wave fits into the length of string between the two fixed ends (Figure 26)—and its higher harmonics. Disturbances of all other wavelengths are forbidden. In a similar fashion, the parallel metal plates allow electromagnetic waves only of a certain wavelength to reflect backward and forward in the gap—a pure "note" of electromagnetic vibration and its overtones. The plate arrangement quashes all of the fluctuations that do not "fit" into the space between the plates.

The upshot of this is that some of the activity of the quantum vacuum is suppressed, and with it some of its energy.

Thinking in terms of virtual photons, there will be *fewer* photons bouncing around in each cubic centimeter of space between the plates than there are in each cubic centimeter outside, because some wavelengths are forbidden. So the overall effect will be of an excess pressure of photons outside the plates, a force trying to squeeze the plates together. Thus, the Casimir effect shows up as a force of attraction between the two plates.

The force is small, but measurable. The virtual photons with short wavelengths are largely unaffected. The longer wavelengths, however, are seriously modified. As longer-wavelength fluctuations are the ones associated with lower-energy quantum disturbances, the total alteration in the vacuum energy is rather small. Nevertheless, it can just be observed, as the force of attraction which Casimir computed. The most satisfactory experiments actually involve curved mica surfaces rather than metal plates, but this is a minor practical detail; the force has been verified quite accurately, including its alteration as the size of the gap is varied. Experiments such as this directly demonstrate the existence of quantum vacuum activity.

The only thing that prevents the virtual particles taking on a real, permanent life is lack of energy. The inherent uncertainty of the quantum world allows them to appear for a short time, without the Universe budgeting for the discrepancy. But the fluctuation cannot be sustained indefinitely, and on longer time-scales the energy books must be kept in balance. Real particles can be created in a similar fashion only by supplying a large enough source of energy. A good example is in accelerator experiments where opposing beams of high-speed protons are made to collide with each other. In an energetic collision between two protons, newly created particles, called pions, are often observed to spew forth. These particles are not pieces broken off from the protons; they have been created out of the energy of motion of the two particles, released when the protons are slowed by the collision. Because no energy has been borrowed from the vacuum, the newly created pions exist as independent, real particles in their own right.

Virtual particles from the vacuum could be directly promoted into permanent reality if enough energy were available. A direct way to do this would be to feed energy into the Casimir plate system by vigorously wiggling one of the plates (this is analogous to plucking the string of a guitar). In fact, just a single moving plate is all that is needed to promote virtual photons into real photons—in principle. As the reflecting surface moves, quantum fields are reflected from it, and if this moving mirror is accelerated, then the virtual photons in the vacuum in front of it are energized, as they bounce off the advancing mirror. In this way, some of the virtual photons should be promoted into reality—a moving mirror should be a *source* of light, not just a reflector. If we could accelerate the mirror fast enough we would literally be able to see the particles of the quantum vacuum with our own eyes.

There is, inevitably, a snag. For a mirror accelerating at one G (with the same acceleration that an object falling in the Earth's gravity would experience) the temperature of the radiation emitted by the mirror would be a mere 4×10^{-20} K. The theory predicts a linear relationship between acceleration and temperature (so that doubling the acceleration doubles the temperature), and visible light has a temperature of about 6,000 K (the temperature of the surface of the Sun, where most of our visible light comes from). Clearly, no ordinary mirror could withstand the sort of treatment needed to produce detectable moving-mirror radiation.

But all is not lost. Researchers at the Bell Laboratories in New Jersey have found a type of mirror that could, in principle, achieve the equivalent of an acceleration in the region of 10^{20} G. It consists of a gas in which the index of refraction is caused to change abruptly, along a rapidly moving plane passing through the gas. One way of doing this is to ionize the gas suddenly using a system of lasers. By carefully manipulating the laser light, a moving front of plasma (gas in which electrons have been stripped from their atoms) might be created that would mimic the effects of a moving mirror, at least at some radiation frequencies. The wave of ionization passing through the gas should act exactly like a moving mirror, and it has been calculated that the phenomenal accelerations required to make the moving-mirror radiation detectable could be achieved. There are problems with the design of such an experiment, including the difficulty of distinguishing the moving-mirror radiation from all the other electromagnetic activity associated with the moving plasma front. Since the idea was only proposed in 1989, experimenters have not yet had time to come to grips with these practical problems.

Another direct way of supplying energy to the vacuum would be to create an intense electric field between metal plates. This would not affect the virtual photons, but it would interact with the virtual electrons, and other electrically charged virtual particles, that inhabit the seemingly empty space between the plates. Given a sufficiently intense electric field, real electrons would appear in the gap as electrical energy supplied their mass.

The electrical energies involved are far greater than anything that could be achieved in a practical experiment involving metal plates. However, it is possible to create fleetingly electric fields of the right strength by smashing together two very heavy atomic nuclei. This momentarily forms a tightly concentrated ball of many dozen protons, each with its own positive electric charge. The combined electric field of this proton ball comes close to the intensity needed to produce pairs of electrons and positrons (the electron's antiparticle) near the surface of the ball. Experiments of this kind have been carried out, and there is some rather ambiguous evidence that such an effect may have been observed.

Although intense electric fields are the most obvious way to boil the vacuum, gravitational fields could also do the trick. Most black holes are probably at least a kilometer across, but it is conceivable that during the big bang, black holes the size of an atomic nucleus could have formed. The *smaller* a black hole is, the more strongly spacetime is distorted in its immediate vicinity (in effect, spacetime has to be wrapped more tightly to surround a smaller black hole). Distorting spacetime strongly implies the presence of strong gravitational fields, and Stephen Hawking has shown that the fierce gravity near such a hole would excite the quantum vacuum to produce real particles, paid for out of the gravitational energy of the hole. Particles would boil off from the hole into the outside world, while the hole itself would lose mass and eventually evaporate completely away in a burst of subatomic debris.

Another example of ultrastrong gravitational fields is the big bang itself. Calculations indicate that during the first 10^{-21} of a second after the beginning, the cosmic conditions were so extreme that prolific spontaneous particle creation would have occurred. This would have involved real particles created out of the gravitational energy of the expanding Universe itself. It is tempting to attribute the origin of all the matter in the Universe to this genesis from empty space—but there is a snag.

The antiworld

A hundred years ago, nobody asked where matter came from. Astronomers generally believed that the Universe had endured for eternity. Even twenty years ago, the standard answer was that the Universe originated in an unexplainable big bang and that all the matter was present at the outset. Today, however, we have the outline of a possible physical explanation for the origin of matter. But for the explanation to work, we also have to explain the nature of *anti*matter, and answer the puzzle of its *absence,* by and large, from the observable Universe.

The idea of antimatter grew out of the two great advances in twentieth-century physics: the theory of relativity, and quantum mechanics. Before the twentieth century, it was generally supposed that matter could not be created or destroyed; that is, the total mass of matter is conserved. But when Einstein presented his special theory of relativity to the world in 1905, the entire concept of mass was transformed. With his famous equation $E = mc^2$, Einstein demonstrated that mass (m) is a form of energy (E); a particle such as an electron can be viewed as a lump of concentrated energy. You get a lot of energy for a tiny mass, because the factor c in the formula is the speed of light, and that, remember, is a very large number (300,000 kilometers a second).

Because energy comes in many forms, matter might be converted into, say, heat energy, in line with Einstein's equation. Support for this speculation comes from the study of the masses of atomic nuclei. A nucleus of oxygen, for example, contains eight protons and eight neutrons. Weigh these sixteen particles separately and you get a total mass about one percent greater than the actual mass of a nucleus of oxygen. The explanation is that this one percent has been converted from nuclear mass into other forms of energy during the assembly of the nucleus. Today we know that it is precisely the energy release from such processes of nucleosynthesis that powers the Sun and stars.

Important though Einstein's work was, it did not immediately suggest that entire particles might disappear (or appear) by exchanging energy with other forms. Protons in a nucleus might weigh less than they do when isolated, but they do not actually vanish altogether. It was the work of Paul Dirac, in the late 1920s, that raised this possibility.

Dirac was interested in combining the new ideas of quantum mechanics with Einstein's special theory of relativity. In spite of its success in explaining the behavior of electrons in atoms, such as their restriction to certain distinct energy levels, the quantum theory developed by Schrödinger, Heisenberg and others in the 1920s did not conform to the principles of Einstein's theory. In particular, the interchange of energy and mass in line with Einstein's equation was not, at first, incorporated into the quantum theory.

The reconciliation of these two great theories was achieved by Dirac in 1929. The centerpiece of Dirac's theory was an alternative equation to the one devised by Schrödinger to describe the motion of electrons in terms of waves. Dirac's equation embodied both the wave nature of electrons and Einstein's ideas about the relativity of motion, with the correct relationship between mass and energy. But there was a subtlety that could not be ignored.

Strictly speaking, the relevant formula relating mass and energy turned out to be not $E = mc^2$ but $E^2 = m^2c^4$. Taking the square root of both sides does indeed give Einstein's familiar equation—but it *also* gives another "root," leading to the alternative equation $E = -mc^2$ (remember that $-1 \times -1 = +1$).

At first, Dirac ignored the second root, because it implied that the energy of the electron could be negative, which seemed meaningless. But its presence as a solution to the equation troubled him, because he could think of no obvious reason why an electron with positive energy should not emit energy in the form of a photon, and thereby make a transition to a state with negative energy.[2] It could then go on making transitions to energy states that were more and more negative, emitting limitless numbers of photons along the way. If this picture were correct, all matter would be unstable.

Then Dirac hit on a solution to the puzzle. The way he got to that solution involved a particular mental image—model—that later turned out to be wrong. But we want to tell the story the way it unfolded in the late 1920s and early 1930s, to show how even imperfect models can be useful in developing an understanding of reality.

A couple of years earlier, Wolfgang Pauli had proposed that some of the observed properties of electrons could be explained if they possess a curious isolationist tendency, which forbids any two of them from getting too close to each other. This is known as "Pauli's exclusion principle," and it is a crucial factor in explaining how the electrons stack up in "shells" at different distances around the nucleus of an atom (like aircraft stacked up and waiting to land at a busy airport), without all collapsing down into the lowest-energy quantum state. Dirac adapted the exclusion principle to the problem of the negative energy states. What, he wondered, if those negative energy states were *already filled* by electrons? Pauli's principle would then ensure that positive energy electrons could not fall down into the negative energy states. But this idea had a bizarre twist to it. We don't actually see all these negative energy electrons. Therefore, reasoned Dirac, they must be invisible.

Although the suggestion that space is filled with a sea of invisible, permanently existing (not virtual) particles might seem fanciful, it led Dirac to make a remarkable prediction.

---

2 It is a well-established general principle that physical systems tend to seek out states of lowest energy.

Suppose, he reasoned, that one of the hypothetical invisible electrons were given enough energy, perhaps by absorbing a photon, to elevate itself to a normal, positive energy state. It would then become visible, and an ordinary electron would appear as if from nowhere. But that would not be all. The elevated electron would leave behind a "hole" in the negative energy sea. Because negative energy electrons are invisible, the *absence* of one must be *visible*. Now, electrons carry negative charge. Which means that the hole—the absence of an invisible negative charge—would appear as the *presence* of a visible *positive* charge. So when a "new" electron appeared in this way, it should be accompanied by something that would seem, to all tests that we could apply, to be a particle with identical mass to the electron but with a positive, instead of a negative, electric charge. This so-called positron is rather like a mirror image of the electron.

Nobody had ever knowingly observed such a particle when Dirac came up with the idea, and since the only positively charged particle then known was the proton, at first Dirac wondered whether the proton might not be the mirror-image counterpart of the electron, even though they have very different masses. But in 1932 the American physicist Carl Anderson found the positron itself. He was studying cosmic rays at the time. These "rays" are in fact very-high-energy particles that bombard the Earth from space, producing all manner of subatomic debris (secondary particles) when they strike the atmosphere. Some of the secondary particles thereby created penetrate to the Earth's surface, where they can leave visible tracks in suitable detectors. One of the tracks studied by Anderson in 1932 was unmistakably produced by a particle with the same mass as an electron, but curving the opposite way in a magnetic field. It could only be due to a positive electron, or positron.

In the years that followed, refinements were made to Dirac's work, and the need to postulate an invisible sea of negative energy electrons was eliminated when it was found that quantum effects would, in fact, prevent positive energy electrons from falling into negative energy states. The image that led Dirac to the notion of antiparticles was flawed. But the truth lay not in the image but in the equations, and the "mirror" solutions to the quantum version of Einstein's equation still remained; so the refined version of the theory still allowed for (indeed, *required*) the existence of individual negative energy electrons—it contained both electrons and positively charged particles. Moreover, it predicted that *every* sort of particle should have an associated mirror image, or *antiparticle*. Thus there should exist antiprotons, antineutrons, and so on, as well as antielectrons (which are still called by their original name of positrons). Collectively, these particles are known as antimatter. After the Second World War antiprotons and other antimatter particles were also discovered in cosmic rays, and now antiparticles of all varieties are routinely made out of energy, along with matter particles, and are even stored by being trapped in magnetic fields, in particle physics laboratories around the world.

Both Dirac and Anderson received Nobel Prizes for their achievements. In his Nobel address, in 1933, Dirac made a further bold proposal. He suggested that it was simply "an accident" that the Earth is made with a preponderance of matter over antimatter. He speculated that "it is quite possible that for some of the stars it is the other way about." In other words, there might be antistars, antiplanets and even, one could imagine, antipeople.

Although scientists have so far witnessed only individual antiparticles, there is indeed no reason in principle why antiprotons and antielectrons could not combine into atoms of antihydrogen. Adding in antineutrons opens up the possibility, as Dirac realized, of heavier elements as well—and of entire antiworlds made up of mirror matter. The physics of mirror matter precisely mirrors that of matter, so that an antistar would look and evolve just like an ordinary star. There would be no direct way in which you could tell from a distance which it was.

On the other hand, when antimatter comes into direct contact with matter, its presence is instantly revealed. The creation of an electron-positron pair out of electromagnetic radiation energy, the process predicted by Dirac, also works in reverse. If an electron encounters a positron, mutual annihilation occurs, with the total mass of the pair of particles being converted into photons. The energy of the photons is so great that they belong to the gamma ray region of the electromagnetic spectrum, more energetic even than x-rays. A similar process, releasing still more energy, occurs when a proton meets an antiproton, or if a neutron meets an antineutron. So any encounter between matter and antimatter results in mutual destruction amid a shower of gamma radiation. For that reason, any antimatter particles produced on Earth, including those from cosmic radiation, are very short-lived.

The fact that particle-antiparticle pairs can be created from energy (it does not even have to be electromagnetic energy) opens the way to an explanation of where the material of the Universe has come from. As we have seen, the big bang triggered processes capable of generating huge quantities of energy, and some of this energy would have gone into creating matter. It is therefore no longer necessary to postulate *ad hoc* that matter was simply present at the outset. Its existence can now be attributed to physical processes occurring in the primeval phase of the cosmos. But this physical process would seem, at first sight, to produce an equal quantity of antimatter along with the matter. The Universe would then be symmetric in matter and antimatter, and there would be as many antistars as stars. There would be a world of antimatter intermingled with that of matter.

The theory that the Universe is symmetric in this way has a pleasing elegance, and it formed the basis of an influential book called *Worlds and Antiworlds*, written by the Swedish astrophysicist Hannes Alfvén and published in the 1960s. Unfortunately, this neatly symmetric theory faces a major obstacle. All cosmological evidence suggests that in its primeval phase the Universe consisted of a "soup" of elementary particles distributed uniformly throughout space. According to the symmetric theory, this soup would have contained equal quantities of particles and antiparticles, mixed together. This explosive mixture would have led to wholesale annihilation, as positrons ran into electrons, protons jostled up against antiprotons, and neutrons encountered antineutrons. Very little material would have been left over.

Some physicists have searched for some mechanism that would lead the antiparticles to congregate with one another, allowing the antimatter to separate from the matter, at least partially. Their goal was to produce separation over regions large enough to contain at least a galactic mass of material. The reasoning behind this was that galaxies are relatively isolated entities, separated by chasms of empty space[3] that would ensure a relatively peaceful coexistence with possible anti-neighbors). But no convincing mechanism to achieve this separation has ever been found.

Meanwhile, astronomical observations have also cast

3 At least, space devoid of everyday matter. Cosmologists now suspect that there may be a different kind of dark stuff in the spaces between galaxies—see Gribbin's *The Omega Point*.

doubt on the existence of large quantities of antimatter in the Universe. These observations involve the detection of gamma rays from space using instruments mounted on satellites. Gamma rays do not penetrate the Earth's atmosphere, but with instruments carried above the atmosphere gamma-ray astronomy became a practical proposition. Satellites have now been used to search for gamma rays from our Milky Way Galaxy and beyond. They have indeed detected gamma rays from the center of the Galaxy with precisely the right energy to correspond to electron-positron annihilation. This was the first direct evidence for the existence of antimatter outside the immediate environment of the Earth. Nevertheless, the amount of gamma radiation from this source and others is so small that estimates limit the fraction of antimatter that might exist in our Galaxy to no more than one part in a million.

Even this figure probably errs on the side of optimism for seekers of antimatter. Interstellar space contains vast quantities of tenuous gas and dust, so that no astronomical object moves in a perfect vacuum. Also, even in space, large objects bump into each other occasionally. As even the slightest brush between matter and antimatter would give rise to copious quantities of distinctive gamma radiation, one can rule out more than a tiny contamination of antimatter in the Milky Way. The best explanation for the characteristic gamma radiation we see coming from the center of the Galaxy is that the positrons involved in the annihilations have themselves been produced relatively recently, by pair-creation events occurring at the energetic heart of our Galaxy; there is no evidence to suggest that the Milky Way contains any antimatter left over from the birth of the Universe.

Similar reasoning can be applied to other galaxies. Even galaxies themselves collide with other galaxies from time to time, and intergalactic collisions must have been much more common in the past history of the expanding Universe, when conditions were more crowded. If these collisions involved encounters between galaxies and antigalaxies, the Universe today would be flooded with gamma radiation left over from the encounters. This simply is not observed. It seems that the vast majority of galaxies are made of matter. We are therefore faced with a conundrum. If the laws of physics are symmetric between matter and antimatter, how has the Universe ended up with a predominance of matter?

Where has all the antimatter gone?

A possible answer to the conundrum emerged from a discovery made in 1964, by two American physicists, Val Fitch and James Cronin, working at Princeton University. They were studying the behavior of a subnuclear particle called the K^0 meson. This is an unstable particle, and "decays" into a mixture of other particles. Fitch and Cronin found that the particle version of the K^0 decays at a very slightly different rate from its antiparticle counterpart. Though the effect is exceedingly small, it is of enormous significance. The long-standing and natural assumption that all physical processes are symmetric between matter and antimatter was, after all, shown to be false. There is actually a tiny asymmetry.

This discovery has an intriguing consequence. Until 1964 it had seemed that there was no way for intelligent beings in different galaxies to determine, by communicating with each other by radio, whether they were both made of the same sort of material, or whether one was made of matter and the other of antimatter. But by measuring the decay rates of K^0 particles in their respective laboratories, and then comparing notes, they could, after all, decide if they were both made of the same sort of stuff—useful information, if they were planning to visit each other and meet physically!

Even more significantly, however, the tiny lopsidedness

between matter and antimatter allows for the possibility of an explanation of why a very slight preponderance of the former over the latter might have emerged from the big bang. It goes like this. In the beginning there was energy, and the energy created particles and antiparticles. Because of the asymmetry discovered by Fitch and Cronin, however, for every billion antiparticles that were created there were a billion *and one* particles. As the Universe cooled, the antiparticles all annihilated with their particle counterparts, leaving just the one-in-a-billion excess of particles to emerge unscathed. These survivors were immersed in a sea of gamma radiation when the Universe was young, a billion or so gamma-ray photons for every particle of matter. As the Universe expanded and cooled further, this gamma radiation also cooled and degenerated into ordinary heat radiation. In fact, the famous cosmic microwave background radiation that still fills the Universe today is a relic of that primeval gamma radiation.

If this scenario is correct, it not only explains how a matter-dominated Universe came to exist, but holds out the promise of an explanation for the temperature of the microwave background radiation as well. This temperature is determined by the average density of photons in space—specifically, how many there are compared with the number of atoms. Hitherto, the ratio of the number of cosmic background photons to the number of atoms in the observable Universe—one of the most important parameters in the whole of cosmology—had been regarded as one of those mysterious numbers that just happened to have the value we observe. From measurements of the microwave background radiation, one finds that the ratio has a numerical value of about a billion to one. This is just the sort of magnitude that now emerges from the calculations of the tiny lopsidedness between matter and antimatter.

If this theory is on the right lines, the presence of matter without antimatter in the Universe today is not the only prediction of cosmological significance, for what can be done can also be undone. The same asymmetry that permitted matter to come into existence from energy, unaccompanied by antimatter, also allows it to disappear again. The theory predicts that this can happen because protons, hitherto regarded as indestructible, are actually very slightly unstable, and will decay into positrons after an enormous duration of time (at least 10^{30} years). If the prediction is correct, it implies that the very substance of the cosmos will inevitably evaporate away, albeit slowly. Since there is one electron for every proton in the Universe, the decay of protons into positrons means that eventually positrons and electrons will meet and annihilate.

The decay itself is a statistical process, like other quantum mechanical processes, and this means that although on average any particular proton will survive for billions upon billions of years without decaying, there is a small probability that out of a large collection of protons—such as a lump of ordinary matter—one or two may decay in any particular year. Experiments have been conducted to search for evidence of such proton decay in huge tanks of water, but so far without success.

If the above scenario is right, it means that any antimatter found in the Universe today would be of a secondary nature, produced as a by-product of high-energy collision events between particles of matter. So long as no direct experimental confirmation of the validity of these ideas exists, however, the question of whether there is any primordial antimatter in the Universe remains open. If it does exist, a good place to look for it would be in cosmic rays.

Instruments carried high into the atmosphere by balloons have detected large numbers of antiprotons in cosmic rays. Most of these are attributed to by-products of collisions between protons deep in interstellar space; but there is a tantalizing puzzle. At low energies, there seem to be far too many antiprotons to fit this explanation. One imaginative suggestion is that these antiprotons are produced by the explosive demise of microscopic black holes by the Hawking process (see Chapter 9). On the other hand, they could represent a trace of primordial antimatter. Nobody can yet be sure of their origin.

A much more definitive discovery, for those seeking *primordial* antimatter, would be a nucleus of some heavier anti-element, such as antihelium. After hydrogen, helium is the most abundant substance in the Universe, so it is reasonable to expect antihelium to be the most abundant antistuff after antiprotons (which are just nuclei of antihydrogen). The point about antihelium is that its nuclei are composite, consisting (in the form that ought to be most common) of two antiprotons and two antineutrons. There is no way that this complicated structure could be created by a random, high-energy collision of particles in interstellar space. Ordinary helium is made by nuclear reactions inside stars today, and was made in large quantities during the late stages of the big bang itself. If even a single nucleus of antihelium were discovered, it would strongly suggest the existence of antistars.

A search for cosmic nuclei of antihelium will be made in the late 1990s by American astronomers using a device called Astromag, which will be installed on the US space station. Astromag will be equipped with powerful superconducting magnets, cooled to near absolute zero, which will be able to bend the paths of electrically charged high-speed particles and antiparticles entering the device. A grid of detectors will be able to discriminate between nuclei of, say, helium and antihelium, which will follow trajectories that bend opposite ways in the magnetic field.

If antistars exist (and we stress that this remains only a very remote possibility), then so, presumably, will a whole range of smaller objects: antiplanets, antiasteroids, anticomets, antirocks, grains of antidust, and so on. A fascinating question then arises. What would happen if something more substantial than the odd nucleus of antimatter should enter the Solar System?

In its passage through the matter-infested regions of the Galaxy, a chunk of antimatter would probably encounter microscopic grains of matter, causing bursts of energy release from its surface. This would tend to fragment the material, and create a cloud of antimatter dust and grains. Some of this debris could then conceivably find its way to Earth. This is a sobering thought. Even a pea-sized pebble of antimatter striking the Earth would cause a kiloton explosion, the equivalent of a small nuclear device, that would scarcely go unnoticed.

Curiously, there was indeed an unexplained mystery explosion, of about this size, that occurred on 13 June 1908, in the remote Tunguska region of Siberia. At first, a large meteorite was thought to have been responsible; but an expedition to the site in 1927 failed to find either a crater or any meteoritic debris, even though there was massive destruction among the trees of the vast forests that cover the region. Numerous theories have been advanced to account for the Tunguska event, ranging from the impact of an icy comet (quite likely) to the passage through the Earth of a microscopic black hole (implausible, if only because there was no similar event to mark the exit of the hole through the other side of the Earth). Willard Libby, who received the Nobel Prize for his invention of the radiocarbon dating technique, has suggested that the Tunguska event might have been caused by a small chunk of antimatter from space. If he is right, this could not be an isolated event, because the presence of one chunk of antimatter in our part of the Universe implies, as we have seen, the presence of other chunks with a variety of sizes. But you can relax. The weight of evidence is generally against the possibility.

The genesis of space and time

The fact that modern particle physics is capable of providing a plausible explanation for the origin of matter is a remarkable achievement. But it falls short of accounting for the origin of the Universe in total, since the Universe consists of more than matter. There are space and time—or space-time—as well. We have seen how the energy needed to create matter can ultimately be traced to the gravitational field of the Universe. But why stop there? Many people would quibble that the creation of matter by gravity is not an example of uncaused genesis; it merely shifts the responsibility on to the gravitational field. We still have to explain where *that* came from. But this question confronts us with a curious twist. Unlike the other forces of nature, gravity is not a field existing within spacetime; it *is* spacetime. The general theory of relativity treats the gravitational field as pure geometry: warps in spacetime. So if gravity created matter, we must say that spacetime itself created matter. The key question then becomes: how did space (strictly speaking, spacetime) come into existence?

Many physicists, even today, balk at the puzzle, and are content to leave the matter to the theologians. But others argue that we must expect gravity, and hence spacetime, to be as much subject to the quantum factor as anything else in nature. In that case, if the spontaneous appearance of particles as a result of quantum effects no longer engenders surprise, why can we not entertain the prospect of the spontaneous appearance of spacetime?

Developing a satisfactory description of this process at work would require a proper mathematical theory of quantum gravity, which is not yet available. Probably such a theory will be achieved only in the context of a synthesis in which gravity is united with the other forces of nature into a single superforce. But we already know enough to sketch out some general features of any such future theory and to see why the achievement of this ultimate synthesis of forces has as yet proved an intractable mathematical problem.

One of the difficulties concerns the scale of quantum gravitational processes. Because it is such a weak force—by far the weakest of the four forces of nature—gravity does not manifest its quantum nature on the scale of an atom or even an atomic nucleus, where quantum properties of the other forces are dramatically apparent, but only on a scale some twenty powers of ten below this, across distances of less than 10^{-33} cm. This tiny distance is known as the Planck length, after Max Planck, the originator of the quantum theory. The associated time-scale, which can be regarded as the fundamental quantum unit of time, is the time it would take light to cross such a tiny distance: 10^{-43} sec, the Planck time. Some physicists believe that at the Planck length spacetime breaks up and takes on features more akin to those of a foam than a smooth continuum. In particular, "bubbles" of "virtual" spacetime will form and vanish again in much the same way that virtual particles come and go in the vacuum.

At the Planck scale, spacetime itself can come into being spontaneously and uncaused, through quantum fluctuations. Each such region of spacetime is only about 10^{-33} cm across, and it generally survives for a mere 10^{-43} sec. More accurately, the concept of time during its fleeting existence is smeared out: there is really no such thing as a shorter time interval than this. Like those virtual particles, the quantum bubbles of virtual spacetime disappear again almost as soon as they are born. At least, this is usually the case. Great excitement among cosmologists has, however, been roused in recent years by the intriguing possibility that such a minute blob of spactime bubbling out of nowhere, a "virtual universe" of unimaginably small size, might have avoided a rapid disappearance and become promoted to the real, astronomically large cosmos we now observe. A suitable promotion mechanism has been identified by theorists, and goes under the name of the inflationary universe scenario.

For the trick to work, somehow the nascent universe has to boost its size from almost nothing to literally cosmic proportions. It must trigger this process very rapidly, during the split second for which a quantum fluctuation is normally allowed to exist. To achieve this extraordinary feat it must somehow avoid the crushing barrier of gravity, which under normal circumstances tries to squeeze the universe back to nothing. What is needed is a titanic *repulsive* force, which can overwhelm the attractive grip of gravity and set the universe on the path of expansion.

In the grip of antigravity

We are led back to the physicists' conception of the vacuum as something much more than "nothing at all." It turns out that it is even possible for the quantum vacuum to become "excited" to a higher level of energy. An excited vacuum would look the same as the true vacuum (that is, apparently devoid of permanent particles), but it would be literally bursting with energy and would survive for only a short while before decaying, thereby releasing its energy in the form of real particles. During its short existence, however, the excited vacuum would possess a very peculiar property: an enormous negative pressure. The concept of negative pressure can be compared to the effect of stretching (as opposed to compressing) a spring—it pulls inward rather than pushes outward. It might be supposed that a universe filled with negative pressure would tend to collapse, under the influence of an inward pull. Curiously, this is not the case. The reason is that pressure as such cannot exert a force; only pressure *differences* do that. Fish that live deep in the ocean, for example, are not crushed by the enormous pressure on them because it is balanced by the equally enormous outward pressure of the fluid in their bodies.

In spite of the absence of a mechanical effect of the huge negative pressure in the excited quantum vacuum, there is an important *gravitational* effect. According to the general theory of relativity, pressure is a source of gravity, in addition to the gravity associated with mass or energy. Under normal circumstances the contribution of pressure to the gravitational field of a material object is negligible. The pressure within the Sun, for example, contributes less than one millionth of its total gravity. In the excited quantum vacuum, however, the pressure is so great that its gravitational effect actually exceeds that of its mass-energy. And because the pressure is negative, the gravitational force it produces is also negative—in effect, *antigravity*. It follows that if a tiny bubble universe were created in a state of excited vacuum (perhaps just by chance, out of many trillions of different bubble universes), then the resulting antigravity would produce just the kind of cosmic repulsion needed to inflate the universe, forcing space to expand with explosive rapidity.

To gain some idea of just how enormous the cosmic repulsion might have been, note that the Universe would have doubled in size in a mere 10^{-35} second during this rapid inflationary phase. And it would have gone on and on doubling every 10^{-35} second so long as it remained seized in the grip of the huge repulsive force. This pattern of growth is known as "exponential," and it leads to a very rapid increase in magnitude.[4] In scarcely more than a billion-trillion-trillionth of a second the Universe would have swelled in volume by a factor of 10^{80}. The region of space visible to us today increased in radius during that tiny time interval from 10^{-26} cm to about 10 cm.

This inflationary phase of frenetic and accelerating expansion lasted for only a very brief duration. The excited vacuum state, being inherently unstable, soon decayed. The enormous reserves of energy locked up in the excited vacuum were, as a result, released in the form of heat and particles of matter. Once the vacuum excitation disappeared, the cosmic repulsive force disappeared, but the momentum of expansion continued, producing the explosive violence that we associate with the big bang. But once the negative pressure had faded, gravity assumed its usual attractive role, acting as a brake on the expansion, a brake that has been operating ever since, eventually slowing it down to the rate observed today.

The significance of the huge and sudden distension represented by inflation is not only that it could convert a minute blob of spontaneously generated spacetime into a veritable universe, but that any initial irregularities present, such as turbulence or an uneven distribution of energy, would be hugely diluted and smoothed away by the colossal stretching

4 The power of successive doublings is graphically illustrated by the old story of how much rice would be needed to meet the requirement that the first square on a chessboard contain one grain, the second two grains, the third four grains, and so on. The last square of the board would correspond to 2^{64} grains of rice, or 18 billion billion grains. Similarly, after only 64 lots of 10^{-35} second (still just under 10^{-33} sec after the beginning) the Universe was 18 billion billion times bigger than it started out, inflated from a size 10^{-33} cm across to a more impressive 10^{-14} cm, about a tenth as big as the nucleus of an atom. In just under another 10^{-33} sec, it would be a kilometer across.

involved. We should expect such a universe to emerge from its inflationary phase with a highly uniform distribution of matter and motion. What does the observational evidence suggest?

As we mentioned in Chapter 4, the background radiation left over from the big bang has remained more or less undisturbed since a very early epoch of the Universe, and so it is a relic that contains an imprint of any primordial cosmic structure. This radiation is astonishingly smooth, varying in intensity by less than one part in at least 10,000. Evidently, the state of the Universe that emerged from the big bang was also highly uniform. Indeed, on a large scale it remains very uniform even today.

In what used to be the conventional model of the big bang, which had no inflationary epoch, this uniformity is a profound mystery. What agency could have orchestrated the primeval explosion in such a way that all parts of the Universe expanded at the same rate everywhere and in every direction? The mystery deepens when account is taken of the existence of horizons in space. As explained in Chapter 4, we cannot see regions of the Universe beyond about 15 billion light-years away because the light from those regions has not yet had time to reach us during the 15 billion years or so since the origin of the Universe. In the past, the region enclosed by this horizon was correspondingly smaller. At one second, for example, it was only a light-second (300,000 km) across.

Extrapolating back to still earlier epochs, at 10^{-35} sec the horizon size was only 10^{-25} cm. Now according to the traditional big bang picture, in which the Universe expands in a smoothly decelerating way, the Universe we see today had a size of about a millimeter at 10^{-35} sec, which is 10^{24} times bigger than the corresponding horizon size. So at 10^{-35} sec the presently observable Universe was divided up, according to this theory, into 10^{72} horizon regions, each "invisible" from all the others. The significance of this is that because no physical force or influence can travel faster than light, regions of space outside each other's horizons can in no way act on each other physically; they are entirely separate and causally independent. How, then, did these separate regions cooperate in their motion in the absence of any communication or force acting between them to impose uniformity?

Inflation solves this horizon problem because of the abrupt swelling in size between 10^{-35} sec and 10^{-32} sec. In the inflationary picture the size of the presently observed Universe was only 10^{-26} cm at 10^{-35} sec, and this is within the horizon region (10^{-25} cm) at that time. So there is no mystery, in the inflationary theory, about the uniformity of the Universe out to the distances we can currently observe.

The solution of the horizon problem is not the only success of the inflationary universe scenario. It also resolves another long-standing enigma concerning the rate of the cosmic expansion. The present expansion is a relic of the explosion that marked the creation of the Universe. In the conventional model, the Universe has been decelerating ever since. Had the bang been less violent, the cumulative gravity of all the cosmic material would have caused the entire Universe to fall back on itself after a brief expansion. Alternatively, had the big bang been bigger, the cosmic material would have been spread more thinly by the stretching of space, and galaxies would never have formed. In fact the vigor of the explosion was matched to the gravitating power of the Universe so precisely that it lies very close to the critical boundary between these alternatives. The theory of relativity provides a connection between the rate of expansion and the average spatial curvature of the Universe. For the critical case of exactly balanced expansion the spatial curvature is zero—space is flat on the large scale.

It is fascinating to compute how finely tuned this cosmic balancing act must be. If we go back to the Planck time, 10^{-43} sec (which is the earliest time we can meaningfully discuss), we find that the matching of explosive vigor to gravitating power was, on the standard model, accurate to no less than one part in 10^{60}. This astonishing fidelity perplexed cosmologists for a long time. Why should the Universe be so propitiously arranged to such phenomenal accuracy?

This is where inflation once again comes to the rescue. *Whatever* the vigor of the initial bang, its effect is utterly swamped by the rising tide of inflation. At the end of the inflationary phase, the Universe has entirely forgotten its original activity, and the behavior stamped on subsequent epochs carries only the imprint of inflation. It happens that exponential inflation delivers to the Universe an expansion rate very close to the critical value, balancing expansion against gravity far more accurately than any human observations can ever hope to measure. A helpful way to understand this result is to imagine an intelligent ant on the surface of a grape. Such a creature might easily determine that the surface of the grape is curved. But if the grape were swelled in size by 64 doublings the ant would never be able to detect the now tiny curvature of the surface it walked upon.

Similarly, inflation can resolve at least part of the puzzle of Mach's principle—explaining why the Universe is not rotating. Any initial rotation would have been reduced to an immeasurably slow, stately progression by the huge early expansion of the Universe, just as the rotation of a spinning ice skater is slowed when the skater's arms are extended.

This catalogue of successes has endeared the inflationary scenario to many cosmologists. The scenario is not, however, without its problems. Foremost among these is the question of how the runaway inflation process comes to an end and returns the Universe to the more conventional activity of gradually slowing expansion. For inflation to work properly, it must be sustained for long enough for the Universe to increase in size at least 10^{25} times. During this period the temperature falls by the same factor, dropping almost to absolute zero. Thus, the Universe almost instantaneously cools from a temperature of about 10^{27} K to nearly zero. The way then lies open for the Universe to drop into its low-temperature phase in which the vacuum assumes the familiar unexcited state it has today and the cosmic repulsion force disappears. This change, which has been likened to the transition from water vapor into liquid water, or water into ice, will obviously bring about the end of inflation by removing its driving force. To avoid this happening too quickly, the original theory, developed in the early 1980s by Alan Guth, of the Massachusetts Institute of Technology, proposed that the cosmic material underwent a period of so-called supercooling.

One example of supercooling occurs when pure liquid water is cooled slowly. It can remain liquid even a little below its freezing point, until a slight disturbance abruptly causes the supercooled liquid to solidify into ice. In a similar way, the high-temperature, excited vacuum phase of the Universe could have persisted for a while after the temperature had dropped away to nothing as a result of inflation, enabling the repulsive force to continue to operate until the necessary amount of swelling had occurred, before the "freeze-out" took place.

Such a phase transition would not occur uniformly throughout the Universe, but by so-called nucleation. Roughly speaking, small bubbles of the new phase appear at random, and grow with the speed of light, eventually intersecting one another and filling all of space. Inside the bubbles inflation comes to a shuddering halt. The energy of the runaway expansion is transferred instead to the bubble walls. When these highly energetic walls collide, they dissipate their energy rapidly in the form of heat, giving back the vast reserves of thermal energy that were sapped from the cosmos during inflation. Thus, the Universe is returned abruptly and explosively to a high-temperature state once more, but this time without the repulsive force. Following this reheating, it is free to continue along the conventional path of decelerating expansion from a hot big bang, with the uniformity, horizon and expansion rate problems already attended to.

Although the broad outlines of this idea are attractive, there are snags hidden in the details, especially concerning the collisions between bubble walls. Such encounters would be random and chaotic, and seem at first sight likely to introduce just the kind of nonuniformity into the universe that inflation was supposed to get rid of. There is as yet no unanimous agreement about the best way to avoid this difficulty, which has become known as the "graceful exit" problem; but several possibilities have been suggested.

One suggestion is that bubbles of the new phase grow to such enormous sizes before collisions between their boundary walls that we live in a region of the Universe beyond the horizon from any such wall, and out of range of any disturbances caused by collisions between walls. Another proposal is that, rather than appealing to supercooling to prolong the inflationary period as required, the phase transition itself may have been a sluggish process.

The essential idea can be illustrated by an analogy. Imagine a ball resting on top of a hill (Figure 27). The system is unstable because a slight disturbance will cause the ball to roll off the hilltop—this is like the excited vacuum state of the Universe. Once the ball starts rolling, it goes all the way

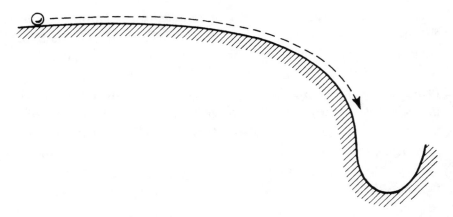

Figure 27. The unstable, excited quantum vacuum state of the early Universe is analogous to a ball poised insecurely on a slope above a valley. If the slope is shallow, the "roll-over" time will be long, enabling the Universe to inflate for an extended period before all the energy is given up as heat.

down into the valley, where it can come to rest in stable equilibrium—this is like the stable vaccum state of the Universe. The top of the hill represents the excited vacuum phase, while the valley represents the everyday vacuum phase. Obviously, the time required for the freeze-out to take place is determined by the time it takes for the ball to roll down the hill. If the slope of the hill were very shallow at the top, the ball would start rolling only very slowly, and this is equivalent to saying that very little change in the nature of the vacuum would occur at first, even though inflation had begun. There are reasons to suspect that the quantum processes that drove the cosmic phase transition did behave in this sort of way, delaying the freeze-out long enough for inflation to work its magic, but avoiding the problems of bubble formation that supercooling would cause, while still giving heat back to the Universe at the end of the phase transition.[5]

The inflationary universe scenario is still in its infancy, and

5 Just as, indeed, water gives up latent heat when it freezes.

new versions continue to be formulated. Many of the details are hard to analyze and are highly dependent on the particular model developed by theorists. It is far too soon to pronounce the theory a complete success. Yet it contains several features that so neatly account for otherwise mysterious cosmological facts that it is hard to resist the proposal that some sort of inflationary activity must indeed have occurred during that first brief flash of existence.

If inflation can be made a complete success, it provides a mechanism for converting a virtual quantum universe into a full-blown cosmos, and enables us to contemplate scientifically the creation *ex nihilo* of theology. A tiny bubble of spacetime pops spontaneously and ghostlike into existence as a result of quantum fluctuations, whereupon inflation seizes it and it swells to macroscopic dimensions. Freeze-out then occurs and the expansion drops in rate amid a burst of heat. The heat energy and gravitational energy of expanding space then produce matter, and the whole assemblage gradually cools and decelerates to the conditions we observe today.

It seems that we have gained something from nothing, contradicting the dictum of the great Roman philosopher Lucretius that "Nothing can come out of nothing." Indeed, as Alan Guth has remarked: "It is often said that there is no such thing as a free lunch; the Universe, however, is the ultimate free lunch." Or is it? All good things come to an end, and in this picture the Universe itself is no exception—its ultimate fate was already sealed in the first second of its existence.

6 . . . And the Last

Perhaps the most fundamental feature of the clockwork universe is the property that once it has been set going, it lumbers on forever, unaided, its destiny already fixed by its past history. In the foregoing chapters we have outlined the new vision of the Universe, one in which the future is open and undecided, in which there is room for spontaneity and novelty and infinite variety. But there is one sense in which the new and old models concur, and that concerns the overall fate of the cosmos as a whole. For although within a limited portion of the Universe the future may not yet be rigidly fixed, when it comes to the totality, the laws of relativity and quantum physics are just as uncompromising as those of Newton. An examination of those laws shows that the manner of demise of the Universe as a whole is determined by the nature of its birth.

As we explained in the previous chapter, the Universe is in the grip of its own gravity, and avoids collapse only because it has been propelled into expansion by an inflationary burst shortly after its birth. The expansion rate, however, is inexorably slowing, and the burning question is whether it will eventually slow to a halt, and turn into a contraction. The issue is closely related to the geometry of space: if space is finite and closed, then the equations of general relativity predict that the Universe will collapse. It is impossible for us to tell by direct observation whether that is indeed the actual state of affairs,[1] but many theorists have argued that it should

---

1 Impossible *in practice*, that is; in principle, superbly accurate observations, the equivalent in three dimensions of drawing triangles on the surface of the

be so on deeper grounds. For example, it is probably only in a spatially closed universe that Mach's principle can be properly formulated. Moreover, Stephen Hawking has proposed a convincing model for the quantum origin of the Universe in which space is required to be closed.

Inflation may have expanded the size of the bubble enormously, but it can never make a closed spacetime open, and in that case gravity must one day win its battle against expansion. This will first halt the growth in the size of the Universe, and then reverse the trend, to produce collapse back toward a tiny volume, and ultimately disappearance into a singularity. It may take a vast stretch of time—perhaps trillions of trillions of years—but in this picture the last second of the existence of the Universe will be a mirror image of the first second, with particles being converted into energy, the energy distorting the fabric of spacetime and ultimately wrapping it around so tight that it pops out of existence. The Universe has only been borrowed from the vacuum, after all; all that inflation has done is to delay the inevitable. In quantum physics, something can come out of nothing for a while, but eventually the debt has to be repaid.

The end of time?

The end of the Universe, along the lines just described, is popularly referred to as the "big crunch" or sometimes as the omega point. It is like a rerun of the big bang in reverse. Instead of the Universe erupting into existence out of nothing, it plunges into annihilation, leaving nothing. And "nothing" here means, literally, nothing—no space, no time, no matter. The big crunch is the complete and total termination of the physical Universe; the omega point is the end of time.

Earth and measuring the sum of the angles, could measure the curvature of space and determine whether the Universe is open or closed.

No scientific prediction has been more momentous than this forewarning of ultimate catastrophe, and it carries with it a prediction of almost equally awesome significance, that all of the matter we can see in the Universe today, all the stars and galaxies put together, represents only something like one percent of the material content of the cosmos.

The prediction is linked to the requirement from gravitational theory (mentioned in Chapter 4) that the Universe must be spatially closed and to the observation that space is very nearly flat. It is straightforward to calculate how much matter there must be for each cubic meter of space to provide the gravity to meet these requirements, and the answer is at least ten and probably a hundred times more matter than we can actually see. And while theorists have uncovered this need for unseen dark matter in order to explain the cosmological structure of the Universe, observers have found a comparable need for dark matter to explain the way in which galaxies move within the cosmos. It is now clear from these studies that both individual galaxies and clusters are held in the gravitational grip of far more dark matter than the stuff we see by the light it emits. Nobody is sure what this invisible stuff is, but the best bet is that it is an unseen residue of exotic subatomic particles left over from the big bang.

Theoretical work suggests that, in addition to creating protons, neutrons and electrons—the particles from which atoms are made—the big bang would have coughed out all sorts of other, more exotic particles. For example, neutrinos, which are so elusive that they could penetrate light-years of solid lead, outnumber protons by about a billion to one. They are relics of the first millisecond. Then there are the so-called axions, photinos and gravitinos, from much earlier cosmic epochs. All these particles interact with ordinary matter so feebly that they have escaped detection so far, although experiments are planned to trap some of them in the near future. Their combined gravity, however, could dominate the Universe and bring about its eventual demise. High-energy particle processes occurring in the first fraction of a second could have produced enough exotic unseen matter to explain where the mass-energy needed to seal the fate of the cosmos is hiding today.

Evidence that some unseen influence is at work in the cosmos comes from studies of the way that galaxies are distributed in space. In stressing the remarkable uniformity of the Universe at large, we have been careful to refer to the *average* conditions over a large enough volume of space, acknowledging that there are irregularities on smaller scales. Although the large-scale uniformity is a key to understanding the initial conditions, the local irregularities are just as important as a guide both to how the Universe began to deviate from uniformity long ago and, it seems, to the ultimate fate of the cosmos. Studies of these irregularities provide insights into both the beginning and end of spacetime, the first and last seconds.

Stringing the Universe together

From a casual glance at the night sky, it is obvious that the stars are not distributed uniformly in space, but are clustered in groups. The broad band of light called the Milky Way is the most conspicuous such clustering. As discussed in Chapter 4, about 100 billion stars, including the Sun, make up the Milky Way Galaxy, a wheel-shaped structure, the visible part of which measures about 100,000 light-years across. Recall that this Galaxy is itself part of a group of galaxies that form a cluster, while the cluster in turn forms part of a supercluster of many thousands of galaxies. Powerful telescopes reveal that this pattern of hierarchical clustering prevails throughout the cosmos.

The origin of this cosmical structure is one of the great mysteries of modern science. Why is matter not spread smoothly across the Universe? What has caused it to aggregate in certain preferred regions of space?

It is tempting to attribute this structure to initial conditions, to claim that the Universe was simply made that way, and that the clustering was imprinted upon it at birth. But this possibility is ruled out by studies of the cosmic background radiation, the heat radiation left over from the big bang. Study of the tiny variations in temperature of the radiation coming from different regions of the sky would reveal irregularities that were present in the hot gas that filled the Universe as early as one million years after the big bang. This epoch, more than 10 billion years in our past, predates the formation of the galaxies. The results of these surveys indicate that the Universe was remarkably smooth at that time, with as yet no clear evidence for any large-scale structure. The very success of the inflationary scenario in explaining why this should be so reinforces the puzzle of how irregularities as large as clusters and super-clusters of galaxies grew up in the Universe after the first million years.

In spite of the smoothness of the cosmic material in the primordial phase, the force of gravity would have caused any minor irregularities that happened to be present to grow steadily, once inflation was over. As soon as a region begins to accumulate an excess of matter, its gravitating power is enhanced, and so it draws in still more matter from its surroundings, in an escalating process. In this way, any initial variations in density would have become amplified. But the inward accretion of matter takes place amid the overall ex-

pansion of the Universe, which opposes it. Although gravity
will cause structure to grow, the rate of growth in the expanding Universe is very slow—too slow, in fact, to explain
the present clustering of matter as arising from purely random fluctuations in density within an initially completely
smooth Universe.

The only way out of this impasse is to assume that something triggered the growth mechanism, that there existed primordial galactic "seeds" around which material could aggregate efficiently. For a long while, cosmologists were resigned
to assuming that the necessary density perturbations were
simply present at the origin—that the Universe was just
"made that way." This does not, of course, constitute an explanation, it just tells us that things are the way they are because they were the way they were. More recently, however,
the possibility has arisen of providing a physical explanation
for the density perturbations, an explanation based on processes that occurred during the first split second. Recall that
the inflationary epoch lasts only so long as the quantum state
of the Universe corresponds to the excited vacuum state.
Once this has decayed to the "normal" vacuum, inflation
ceases. But the decay process, like all quantum processes, is
subject to fluctuations, in line with Heisenberg's uncertainty
principle. As a result, inflation would not have ceased everywhere simultaneously; some regions of the Universe would
have stopped inflating before others. The upshot of this variation is thus to cause density irregularities in the postinfla-
tionary Universe. So inflation has the remarkable double effect both of wiping out any preexisting irregularities and of
imprinting its own irregularities on the Universe. Moreover, it
turns out that these quantum irregularities have the right
general form to explain the large-scale structure we see
today. If this theory is a good description of the real Uni-

verse, it means that microscopic quantum fluctuations, which originated with quantum uncertainty, are to be seen imprinted across the sky—that the galaxies are relics of "frozen" fluctuations from an epoch no later than 10^{-32} sec.

In spite of its appeal, the quantum fluctuation theory is not without problems. For example, many calculations indicate that the density variations would actually be too large to match the observed irregularity in the Universe today, and there are other, technical difficulties that make the idea less attractive. There is, however, a contending theory which also attempts to explain what triggered the growth of galaxies. It, too, appeals to the very early stages of the Universe when the quantum vacuum decayed from its excited phase. The theory draws a close analogy between this phase transition and the more familiar sort, such as the onset of ferromagnetism. If an iron magnet is heated above a certain critical temperature, called the Curie point, it loses its magnetism. When it cools, the iron undergoes an abrupt transition back to its magnetic phase. This phase transition does not usually, though, take place in the same way everywhere throughout the iron. Instead, the iron becomes divided into distinct domains, each of which has a magnetic field oriented in a particular direction. In the same way, it has been suggested that the cooling Universe would have had a domain structure in which the fields associated with the various forces of nature would have assumed different forms.

The situation at the boundaries between these domains is interesting, because at these locations there will generally be a mismatch in the arrangements of the fields on either side of the boundary. The result could be a kind of dislocation, and under some circumstances the fields could end up being tied in a sort of knot. We described how this type of topological defect can arise in Chapter 2. One feature that would be produced by these mismatches is a series of slender tubes. Outside such a tube there would be the usual empty space corresponding to the normal quantum vacuum we observe today; but inside the tube the quantum state would remain trapped in its excited primordial phase, unable to undergo the phase transition that occurred elsewhere during the first split second. The result is an object known as a cosmic string. If they exist, cosmic strings are time capsules left over from the creation event. They are not made of matter; they are essentially tubes of field energy. Within these tubes the Universe remains frozen in the state it possessed a mere 10^{-35} sec after the beginning.

Cosmic strings are hypothesized to have some bizarre properties. In the most favored version of the theory, the strings are forbidden to have ends, so they must either be infinitely long and stretch right across the Universe, or form closed loops. So concentrated is the field energy within the strings that a kilometer length might typically weigh the same as the Earth. But the significance of this really becomes clear only when you discover how thin that string would be. It is unimaginably thin—a mere million-trillion-trillionth of a centimeter across. One way of grasping what this means is to imagine a cosmic string that stretches across the visible Universe, 10 billion light-years in length. Such a string could be wound up into a ball inside a single atom, and still leave room to spare. This subatomic ball of string would then weigh some 10^{44} tons—as much as a supercluster of galaxies!

Another bizarre property of cosmic strings is that in spite of their enormous mass per unit length, a straight string would normally exert no gravitational force at all on a nearby object. The reason for this weird property can be traced to the internal nature of the string. As we have explained, cosmic strings are essentially energized empty space, tubes into which certain fields of force are squeezed. As well as possessing huge energy (and therefore gravity), these fields also exert an equally huge pressure. The distinctive feature of this particular field pressure—it is the same primeval field pressure that drives inflation, but now trapped within the tube—is that it is negative, so that the field tries to pull rather than push. This means that the strings suffer an enormous tension that tries to shrink them. It also means, as we explained in describing the inflationary force, that there is antigravity associated with the negative pressure in the string. In a straight string, the antigravity of the pressure exactly cancels the gravity of the energy in the tube, leaving zero net gravitational force outside.

This does not mean, however, that the string produces no gravitational effect of any sort. It certainly does. Although the string does not produce a curvature of spacetime around itself, it does produce a distinct effect on the geometry of space, an effect that can be envisaged as follows. Imagine an observer who took a circular trip around such a string. You would expect on the basis of everyday experience that the journey round the circle would involve turning through 360°. If such an observer measured how much turning the trip actually involved, though, he would find that the answer was *less* than 360°.

A helpful way of understanding this is to imagine cutting a wedge out of a flat disc, and gluing the edges of the gap in the disc together (Figure 28). The flat piece of paper has now become cone-shaped, and although the circumference of the former disc is still a circle, it is shorter than it was before. In the case of a cosmic string, the sheet of paper represents a section through space perpendicular to the string, with the apex of the cone corresponding to the point of intersection of the sheet with the string. The effect of the

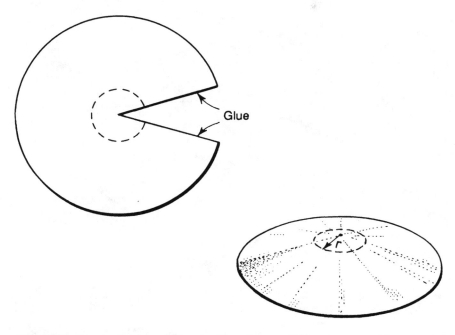

Figure 28. If a wedge is cut out of a flat disc, and the edges of the wedge are glued together, the disc becomes a cone. the cone has the property that the length of a circle centered on the apex is less than 2π. The space in a direction perpendicular to a straight cosmic string would have such "conic" geometry.

string is to cut a wedge out of space, to make space "conical."

The missing-angle property leads to some distinctive effects. For example, two light rays that start out parallel and pass either side of the string can be deflected so that they intersect. The string therefore acts a bit like a cylindrical lens. If a string lies between the observer and some object such as a galaxy or quasar, then the observer may see a double image of the object (Figure 29). For typical strings, the angular displacement is only a few seconds of arc; interestingly, though, astronomers have discovered many close pairs of quasars in

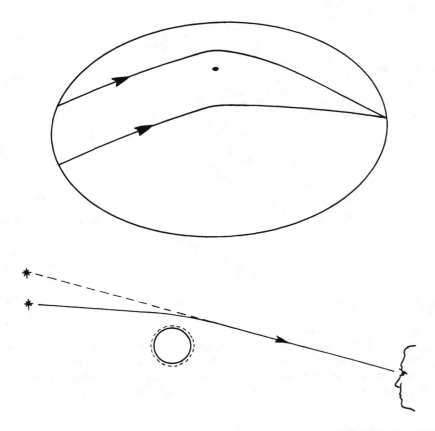

Figure 29. Light rays that start out parallel moving in a conical space can be focused, as if by a lens. An observer may therefore see two images of a source of light on the other side of a cosmic string, not one.

which the spectrum of the light from the two components is identical, and they have concluded that these must be two images of the same object seen through some sort of gravitational lens. Unfortunately, double images could also be produced by the distortion in space caused by the gravity of an intervening galaxy or a black hole, so there is no proof that cosmic strings are responsible. A more careful study of such

multiple images could, however, eventually discriminate between the effects of strings and the influence of galaxies. For
example, if a string lies across our line of sight to an extended object, such as a galaxy, then one of the images we
see should be truncated, with a sharp edge.

Further observable effects occur when one takes into account the motion of the string across the line of sight. The
light from distant astronomical objects is shifted by the expansion of the Universe toward the red end of the spectrum,
and the amount of this red shift provides a measure of the
relative speed with which the object is receding from us. If a
string passes between us and the object, there will be a sudden change in the red shift we observe. A similar effect is
produced on the cosmic background heat radiation: string
motions produce abrupt changes in the temperature on either side of the string, as seen from Earth. Such temperature
variations could be detectable in the near future.

Although a straight string exerts no gravitational force,
from a distance a loop of string behaves more or less like an
ordinary blob of matter. Some cosmologists feel that such
string loops may constitute the "seeds" that triggered the
growth of galaxies and other large-scale structures in the
Universe. But would one expect so many string loops to
have formed in the early Universe? According to mathematical analysis, large numbers of strings would have formed,
moving randomly at close to the speed of light. Clearly, this
would have produced a complicated entanglement as strings
intersected each other. When two strings intersect, the fields
inside them usually interact in such a way as to cause the
two tubes to reconnect so that each end of one string joins
to its erstwhile neighbor (Figure 30). This means that repeated interactions between strings, especially with strings
that tangle and cross over themselves, tend to produce

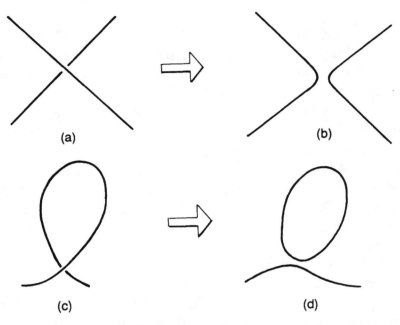

Figure 30. When cosmic strings intersect, they tend to reconnect as indicated, with the strands swapped over. This means that a self-interacting string will often throw off a disconnected closed loop.

closed loops. It seems likely that during the first one second of the Universe there was indeed a profusion of such loops.

Over the eons of time that followed, the Universe expanded enormously. The loops would have moved apart, and also slowed down until they were more or less at rest relative to the gaseous cosmic material. Then, in this more quiescent state, they would have begun accreting material to form galaxies. Many theorists are convinced that cosmic strings played a key role in structuring the Universe on a large scale, and that some strings remain in the Universe at the present epoch. If so, the question arises of whether they can be detected somehow. One possibility has already been mentioned—the curious double-imaging effect. But where should we look for it?

At first it might seem that a good place to search for a loop of cosmic string would be at the center of a galaxy, such as our own Milky Way. But not all of the primordial loops—indeed, not many—will have survived. The fate of a loop depends upon its dynamics. The string tension will try to make the loop shrink, but this will be opposed by the momentum of various string sections that might be moving very rapidly. Computer simulations suggest that, when first formed, a loop will be writhing about wildly. This wriggling motion will lead to a rapid loss of energy from the loop, in a rather exotic form of radiation known as gravitational waves.

Ripples in space

A massive object like the Sun produces a warping of space-time in its vicinity. As the Sun moves, the space warp and time warp move with it. In the depths of the Universe, other objects, some much more massive than the Sun, carry their own space and time warps. If two objects collide, their space and time warps are disrupted, and can release ripples into the surrounding Universe. These ripples are gravitational waves.

The existence of gravitational waves was predicted by Einstein in 1916, but in spite of several decades of experimental effort no such waves have so far been detected. Yet physicists are very confident that they exist. Their elusive nature is due to the feebleness of gravity, so that even powerful gravitational waves could pass right through you without producing a noticeable effect.

It is not only the collision of objects that produces gravitational waves. In theory, most moving masses should emit some gravitational radiation. Common sources in the Universe might include exploding or collapsing stars, the orbital motion of binary stars or the wiggling of a cosmic string. The radiation that is released in such processes travels at the speed of light, and could in principle reach us from the edges of the observable Universe.

So how might gravitational waves be detected? A radio wave manifests itself by wiggling electrically charged particles (such as the electrons in the metal of a radio mast) back and forth; but because gravity acts on everything, not just charged particles, a gravitational radiation detector could in principle be made of anything at all. Unfortunately, because gravity is such an extremely weak force, matter is almost transparent to gravity waves. Most of them pass right through the Earth unchecked. Detectors of unprecedented sensitivity will be required if their passage is ever to be recorded.

Such detectors are now being designed and built. The original detector, built by Joseph Weber, of the University of Maryland, in the 1960s, consisted of an aluminum cylinder 1.5 meters long suspended by a thin wire inside a vacuum chamber. The passage of a gravity wave will have the effect of causing a tiny vibration in the cylinder. Delicate electrical sensors were glued to the cylinder to record such movements. But they are *very* small; success depends on being able to detect changes in length that are truly mind-boggling in their smallness. For example, to be sure of achieving the sensitivity necessary to detect bursts of gravitational waves on, say, a once-a-month basis, changes in bar length of a mere 10^{-20} cm must be discerned. This is equivalent to measuring a change in the Earth's distance from the Sun of only the size of the diameter of a single atom, or a change in the distance to the nearest star ($4\frac{1}{3}$ light-years away) of less than the breadth of a single human hair. It seems impossible, but the experimenters say otherwise. Not only does this require very sophisticated measuring techniques, but there is also the problem of other motions (including seismic vibrations, and even the jiggling motions caused by heat in the cylinder) swamping the effects of gravitational waves. All these extraneous vibrations have to be suppressed.

The physics community was agog when, in the early 1970s, Weber produced evidence of frequent bursts of activity, which he attributed to gravitational disturbances passing through his equipment. Other researchers rushed to develop similar detectors, but without any success in finding gravitational radiation. Since then, detectors have been cooled to near absolute zero to reduce thermal noise, and the sensitivity has been stepped up, but still no independent evidence has been forthcoming. Most experimenters have concluded that the vibrations detected by Weber some twenty years ago were not, in fact, caused by gravitational waves.

Meanwhile, some research groups have turned to alternative detector designs. One of the most promising approaches uses lasers, with precision-tuned beams of light bounced off an arrangement of mirrors, delicately suspended in evacuated tubes so that air currents cannot disturb them. As a space ripple sweeps by, minute changes in the distances between the mirrors will occur, which can, in principle, be detected by careful comparison of the laser light reflected back and forth between the mirrors. Although experimenters continue to make good progress in developing the ultrasensitive techniques required, and in isolating detectors from everyday disturbances, it is likely to be some years yet before there is definitive detection of the long-sought gravity waves. But the confidence of the experimenters that their work will not be in vain has been boosted by the discovery, made by a team at the University of Massachusetts in Amherst, of the effects of gravity waves at work in space. Using the huge radio tele-scope at Arecibo in Puerto Rico, the team has for several years been studying the motion of an unusual system known as PSR 1913 + 16.

This object is a binary star system—two stars in orbit around each other. But it is a binary system with a differ-ence. *Both* stars have collapsed into balls no larger than a terrestrial city, in spite of the fact that each star contains more material than our Sun. The effect of this enormous compression is to raise the density of material in the stars to a colossal level. A teaspoonful of matter there would contain a billion tons! Under these conditions even atoms are crushed, so that these collapsed stars are composed mainly of neutrons.

Neutron stars are thought to form during supernova explo-sions, when the core of a massive star implodes under its own weight. When first formed, they probably spin at a fan-tastic rate, perhaps several hundred times a second. Most stars possess magnetic fields, and if a star collapses, the field is squeezed and intensely amplified. A typical neutron star will, as a result, have a magnetic field a trillion times stronger than that of the Earth. When the star rotates, the magnetic field rotates with it, and the awesome object becomes a powerful cosmic electrical dynamo. Charged particles, such as electrons, in the vicinity of the neutron star get caught up in the magnetic field and whirled around at nearly the speed of light. This forces them to emit strong beams of electromagnetic radiation, including both light and radio waves. As the star spins, so the radiation beam sweeps around, rather like the beam from a lighthouse. For an observer on distant Earth, the effect is to cause a sud-den pulse of light or radio waves each time the beam flicks across our planet.

Rhythmic radio pulses of this kind were first discovered in the late 1960s. Many such objects are now known, and have been dubbed pulsars. But PSR 1913 + 16 is one of only a handful of systems in which a pulsar is in orbit around another neutron star; for this reason, it is known as a "binary pulsar."

This fortuitous arrangement provides a rare opportunity to see the effects of gravitational radiation at work. The orbital period of this binary pulsar—the time it takes for one star to orbit once around its companion—is only about eight hours, which means that the stars are moving at high speed in an intense gravitational field. The cavorting stars are therefore a strong source of gravitational waves, and as the waves flow away into space they deplete the system of energy. As a result, the orbit slowly decays, and the neutron stars spiral toward each other. Eventually, they will collide and coalesce. Meanwhile, the regular radio pulses from this system provide an ideal means to monitor the orbital decay. In effect, the pulsar is a fantastically accurate clock, and as the "clock" moves about in the gravitational field of its companion, so the radio blips vary slightly, due to the effects of gravitation on time. By monitoring the pulses over several years, astronomers have been able to construct an accurate mathematical description of the orbit. When this binary pulsar showed unmistakable signs of orbital decay, scientists became very excited, since it was possible, for the first time, to test Einstein's decades-old prediction that such a system ought to be producing gravitational radiation—a prediction made before anyone knew that neutron stars existed. It is now known that the rate of orbital decay in the binary pulsar matches perfectly the prediction stemming from Einstein's general theory of relativity. It seems clear that, even if gravitational waves have not yet been detected on Earth, we

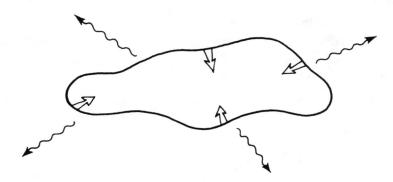

Figure 31. A wriggling loop of cosmic string is a prolific source of gravitational waves. As the waves flow away from the string, so they take energy away, and the loop must shrink as a result.

are at least witnessing their emission elsewhere in the Galaxy.[2]

Just as whirling neutron stars produce gravitational radiation, so should a moving cosmic string (see Figure 31). In the case of a wriggling loop of string the gravitational waves will flow away into space, and cause two effects, one of them rather dramatic. The emission of radiation will not occur uniformly around the loop, but will tend to be strongly beamed in certain directions, depending on the shape and speed of various loop fragments. The momentum carried away by the radiation will cause a reaction on the loop, accelerating it in the opposite direction, rather like a rocket. It has been estimated that, as a result of the rocket effect, string loops could eventually be propelled to as much as 10 per-

2 One word of caution. Scientists also use the term "gravity waves" to describe a wave pattern produced by a fluid, such as the oceans or the atmosphere of the Earth (or the water in your bath), slopping about in a gravitational field; although the name is the same, the phenomenon being described is completely different. If you ever see a headline or story referring to the measurement of gravity waves in the Earth's atmosphere, it does *not* refer to another triumph of general relativity!

cent of the speed of light. Thus, if such loops did constitute
the seeds of galaxies, it is likely that they will have escaped
long ago from the centers of galaxies by this process.

The second effect of the loss of energy by gravitational radiation is to damp out the wriggling motion of the loop, allowing the tension in the string to make it shrink. Ultimately,
a contracting loop of string would shrink away entirely or
perhaps become a black hole. One way or the other, it is unlikely that many loops have survived at the centers of galaxies until the present epoch.

The cumulative effect of gravitational waves emitted from
myriads of primordial string loops would be to fill all of
space with a jumble of ripples, rather like the surface of a
pond agitated by gusts of wind. These gravitational waves
would have an enormous range of wavelengths, some of
them with many light-years between "peak" and "trough,"
reflecting the large initial sizes of some of the loops. Among
other things, the effect of this background of ripples in space
will be to disturb the regularity of pulsar pulses—not, this
time, because of any gravitational wave emission from the
pulsars themselves, but because the space between a given
pulsar and the Earth would be rippling while the pulse was
passing through on its way to us.

The more rapid a pulsar is, the more sensitive its pulses
will be to this effect. Some pulsars spin so fast that their
pulses occur at about a thousand times a second—successive
pulses are only a little more than a millisecond apart. These
"millisecond pulsars" are now being carefully observed for
the telltale signs of any passing gravitational waves left over
from an earlier era of prolific wriggling string loops.

Awesome encounters: cosmic string meets black hole
Because a cosmic string is absolutely forbidden to snap, an
intriguing question occurs concerning the fate of a string that

meets a black hole. Anything that enters a black hole, including a section of cosmic string, can never escape again, yet the black hole cannot bite a piece out of the string without leaving two free ends. The only solution is for the string to become permanently attached to the hole. When this happens the hole starts to suck in the string like two strands of spaghetti. In the case of a straight string passing through a black hole there are no markers on the string to provide a measure of the rate of ingestion. As far as an observer is concerned, nothing would seem to be happening. Indeed, the situation remains static: the black hole does *not* grow in size as a result of gobbling the string. The reason for this is the same as the reason why a straight string doesn't exert any gravitational attraction—the antigravity of the pressure inside precisely cancels the gravity associated with the energy. Consequently there is no net change in the gravity of the black hole as a result of eating a straight string segment, no matter how long that piece of string may be.

In a realistic case, however, the capture of a string by a black hole would be a complicated affair, and the string would certainly not be perfectly straight. Computer simulations carried out by Ian Moss and his colleagues at the University of Newcastle upon Tyne show that when a piece of the string approaches the hole it first develops a sharp kink, or cusp, that points toward the hole. This kink then turns into an open loop, like one coil of a spring, which may then develop another kink, then a loop within the loop, and so on. By the time the string enters the hole, it resembles not so much a single strand of pasta to be slurped up, but a bowl of carelessly served spaghetti strands. If the hole is rotating (as presumably most are) the "spaghetti strands" will also be twirled around, adding to the complexity.

Interest in encounters between cosmic strings and black

holes extends beyond such possible astrophysical scenarios to the fundamentals of physics. One of the key properties of black holes, first investigated by Stephen Hawking, is that they cannot decrease in size. More precisely, the surface area of a black hole can only increase or remain the same. The only exception to this law concerns microscopic black holes, the ones for which quantum processes can convert gravitational energy into real particles, leading to the evaporation and eventual disappearance of the hole in an explosion of energy.

The area increase law is central to the consistency of physics, because it enables the laws of thermodynamics to be applied to black holes. The area of a black hole provides a measure of its entropy, and if a black hole could shrink, that would amount to a decrease in entropy, in violation of one of the most sacrosanct laws of science.

At first sight, it does seem as if the area of a black hole would be decreased if a string passed into it. The reason has to do with the way a string changes the geometry of space-time in its immediate vicinity by cutting out a wedge (recall Figure 28). In the same manner a straight string passing into a black hole would cut a wedge from the hole's surface, thus reducing the surface area of the hole in apparent contradiction of the fundamental theorem we have mentioned. Or will it? In spite of appearances, most theorists are confident that the area law, and the principles of thermodynamics that it relates to, will still be obeyed. One conjecture is that as the string falls into the hole it must deliver some energy which raises the mass of the hole, and hence its radius. The supposition is that this will always increase the surface area more than enough to compensate for the area lost in the wedge cut out by the string.

Before we leave the topic of cosmic strings, we should

mention that their formation involves physical processes that occur at more or less the same epoch as inflation. The crucial question is whether it is just more or just less. Clearly, if strings formed prior to inflation, they would be inflated away like any other irregularities—that is, the density of strings in the Universe would be practically nothing after inflation, and there would be scant hope of finding even a single string within the visible Universe. For this reason, the inflationary scenario and the cosmic string theory are often considered to be incompatible alternatives. This has not, however, prevented some theorists from striving to contrive a mechanism to have both.

Like many of the ideas discussed in this chapter, such attempts by theorists to have their cake and eat it too depend on calculations involving quantum physics. So far, we have avoided treating this topic in detail, because it has a reputation for being subtle and obscure. Some of its predictions are also pretty weird. To develop our story further, however, we shall need to digress somewhat into the details of the quantum theory, and this constitutes the subject of the next chapter.

7 Quantum Weirdness

Whenever you glance at a luminous clock, you are witness to one of the most peculiar processes in nature. The luminosity is caused by a form of radioactivity known as alpha decay, and from the earliest days of its discovery at the end of the nineteenth century it was clear that alpha decay is a very odd phenomenon.

The New Zealand physicist Ernest Rutherford was one of the first to experiment carefully with alpha "rays," as they were then known, and gave them their name in 1898. By 1907, Rutherford and his colleagues had established that alpha particles are, in fact, helium atoms from which two electrons have been removed; such a stripped-down atom later became known as a nucleus, and we now know that each alpha particle is made up of two protons and two neutrons. But it was only a couple of years after the identification of alpha particles with stripped-down helium atoms that Rutherford and his colleagues discovered the basic structure of the atom, using alpha particles as tiny projectiles.

In these experiments, beams of alpha particles were fired at thin sheets of gold foil. Most of the particles went through the foil like "an artillery shell through tissue paper," in Rutherford's words, but just a few particles were deflected by large angles, as if the "artillery shell" had bounced off something solid. Rutherford realized that this could be explained if most of the mass of the atom was concentrated in a compact nucleus. He suggested that each atom consisted of a swarm of very lightweight electrons surrounding the nucleus in a very diffuse cloud. The atom thus resembles, in some respects, the Solar System, in which relatively lightweight planets orbit around a central concentration of mass, the Sun; Rutherford's proposal therefore became known as the planetary model. In place of gravity, however, the atom is held together by electrical forces. Each electron carries one unit of negative charge, while the nucleus carries a total positive charge equal to the number of electrons in the cloud outside. If this structure correctly describes the atom, argued Rutherford, then alpha particles would pass straight through the electron cloud, brushing the electrons aside without being deflected much; only alpha particles that fortuitously happened to hit a nucleus almost head-on would be knocked to one side.

But now Rutherford was confronted by a mystery. If alpha particles are fragments ejected from, say, uranium nuclei, then there must exist a mechanism for each of them to get out of its parent nucleus. Once outside the positively charged nucleus, the positively charged alpha particle would naturally be repelled, and leave the atom altogether. On the other hand, when these same alpha particles are directed toward other uranium nuclei, they just bounce off, with their positive charge repelled by the positive charge of the nucleus. Why is it, wondered Rutherford, that positively charged particles can be held together in the nucleus, and yet the alpha particles don't go back into the uranium nuclei? Surely if they are able to get out, then they should be able to get back in again?

In the 1920s, physicists developed the idea that the charged particles in the nucleus are held together by a strong nuclear force, which overwhelms the electric force at short distances. Together, the effect of the short-range nuclear force and the long-range (but weaker) electric force is to produce an invisible barrier around the nucleus. An alpha particle inside the nucleus is trapped by the barrier, while one approaching from the outside cannot get in. A useful way to imagine this is to think of the nucleus as a collection of particles inside the crater of an extinct volcano. If they can get enough energy to climb the wall of the crater, they can roll over the top and escape; similarly, a particle coming in from outside has to climb the mountain before it can fall into the crater. But this still left the puzzle of why particles that could *escape* from the nucleus could not *get back in* from outside. Detailed calculations of the nature of the barrier only deepened the mystery. It turned out that the particles that were seen to be escaping from the uranium nuclei did not actually have enough energy to surmount the barrier in the first place, yet experiments showed that even particles with twice as much energy did not penetrate the barrier from the outside. It was as if the alpha particles were somehow escaping by tunneling through the barrier. Clearly, something very weird was going on. That weird something—the tunnel effect—was explained by the Russian-born physicist George Gamow in 1928, drawing on the new theory of quantum mechanics, which had been developed largely in response to a whole range of puzzles concerning the atom.

The quantum tunnel

When Rutherford established the basic "planetary" architecture of the atom he had no idea how the electrons could remain in stable orbits around the nucleus. There was, in fact, a deep mystery surrounding the stability of this structure, since the laws of classical mechanics and electromagnetism require that orbiting charged particles should continuously radiate away energy in the form of electromagnetic waves, and spiral down into the nucleus as a result. In other words, according to classical theory the atom should collapse. The actual situation is quite different. The electrons are found to occupy only certain fixed energy levels, equivalent to orbits at different fixed distances from the nucleus (Figure 32). Electromagnetic radiation certainly can be emitted from atoms, but only in discrete and sudden bursts. When this happens an electron hops down abruptly from one level to another.

The reason for the existence of distinct atomic energy levels was a vexing puzzle. How did they arise? What kept the electrons in them? The Danish physicist Niels Bohr took up the problem after visiting Rutherford, who was at that time working at the University of Manchester, in 1912. Bohr's insight led him to propose a mathematical formula that correctly gave the energy levels of the simplest atom, hydrogen, and which stipulated the amount of energy released or absorbed by an electron in jumping between energy levels. This was hailed as a great advance; but nobody knew why the formula worked.

A key feature of Bohr's formula is the appearance of a quantity called Planck's constant, which had been introduced at the turn of the century by the German physicist Max Planck in order to explain the nature of heat radiation. Planck's constant had also been used by Einstein in 1905 to explain the photoelectric effect, a process in which light falling onto a sensitive surface produces a flow of electricity. The work of Planck and Einstein had established that heat radiation and light (and, indeed, all forms of electromagnetic radiation) could not simply be explained in terms of electromagnetic *waves,* but would also behave, under some circumstances, like a stream of particles, now called photons. Planck's constant defined the amount of energy carried by each photon associated with a particular wavelength of radiation. The photons are like little packets of energy—quanta,

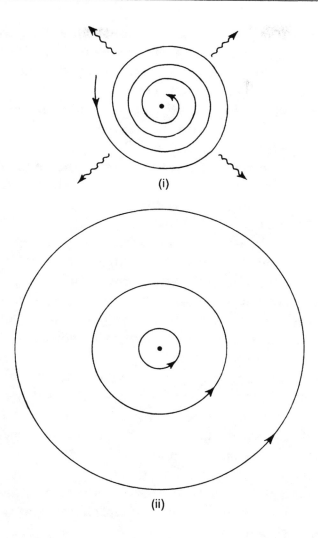

Figure 32. (i) According to classical physics, an electron in orbit around an atomic nucleus should emit electromagnetic radiation continuously, because it is continually accelerating in a curved path. The resulting loss of energy implies that the electron should spiral into the nucleus in a very short time.

(ii) Niels Bohr proposed that atomic electrons are restricted to certain fixed (quantized) orbits. An electron can jump suddenly between these orbits by absorbing or emitting a photon with the appropriate precise wavelength.

as they became known. By demonstrating the need for
Planck's constant in his formula for energy levels, Bohr thus
established a link between electromagnetic quanta and
atomic structure. The discrete energy levels on which electrons were allowed to sit also depend, like the energy of
photons, on a formula involving Planck's constant.

But the mystery concerning why the electrons' energies are
"quantized" to certain discrete levels still remained. The be-
ginnings of an answer came in 1924. A French student, Louis
de Broglie, came up with a bold idea. If light waves can behave like particles, perhaps electrons, which everyone then
thought of as material particles, could also behave like
waves? Developing this idea, de Broglie produced a simple
formula showing how the wavelength of such a particle
might be related to the momentum of the particle. Momentum is the product of mass and velocity; de Broglie suggested that converting momentum into wavelength involved,
once again, Planck's constant.

Although de Broglie did not provide a detailed theory of
matter waves (that was achieved by the Austrian physicist
Erwin Schrödinger), his idea provided a graphic model for
the way that electrons might occupy only certain energy levels around the nucleus of an atom. If an electron is in some
sense a wave, then in order to make the wave "fit" into an
orbit around the nucleus, the size of the orbit must correspond to a whole number of wavelengths, so that when the
wave is wrapped around the nucleus it will join together
smoothly. Only certain discrete energy levels are allowed because only at certain distances from the nucleus will the
wave patterns join up consistently.

The details of this wave fitting were supplied by Schrödinger, who produced an equation that described how electron waves behave in the vicinity of an atomic nucleus. With
the solution of Schrödinger's equation, Bohr's formula for the

energy levels of the hydrogen atom reemerged. This was a major triumph for physical theory, and the starting point of a new era of physics. In the following years the new theory—called quantum mechanics—was applied successfully to many problems involving electrons. Schrödinger's equation now forms the basis of all atomic, molecular and solid-state physics, and much of physical chemistry. But such sweeping success came at a price. As Schrödinger himself was well aware (and as the name of the new theory implies), it rested upon throwing out Newton's time-honored laws of mechanics and replacing them by the new equation for matter waves.

If electrons can behave like waves, then it seemed reasonable to expect all subatomic particles to do so, and this was soon confirmed by experiment. Once the wave nature of subatomic particles had become established, it became clear that some peculiar things could happen on the scale of atoms and nuclei. Suppose, for example, that a stream of electrons encounters a field of force such as an electric barrier. If the force is repulsive, it seems, on the basis of our everyday experience, that the electrons should be deflected away from it. On the other hand, if the force is attractive, we would expect the electrons to be pulled toward the force. Viewed in terms of waves, however, this naive expectation is confounded. Rather like the way in which a pane of glass in a window will reflect some light while allowing most to pass through (so that you can see a ghostly reflection of yourself in the window), so an attractive force field will always reflect some waves. That means that some electrons, out of a large number in a stream, will sometimes bounce back from a region of attraction. It is as though a golf ball, rolling toward the hole, were to reach the brink of the hole and then suddenly reverse direction, instead of falling in.

Given such weird behavior, it is not hard to find an explanation for the mystery of how an alpha particle can tunnel out through the nuclear force barrier. Just as an electron has an associated wave, so does an alpha particle. We must envisage this alpha particle wave as confined within the nucleus by the strong force barrier. Confinement occurs because where an outward-moving wave encounters the barrier it reflects back into the interior of the nucleus. The wave is trapped in the same way that light might be trapped within a box lined with mirrors.

When a light wave reflects from a mirror, however, it does not simply bounce off the reflective surface. In ordinary "glass" mirrors, the reflecting surface is actually a thin film of metal coating the back of the glass. The light wave, as it is being reflected, creates a disturbance that penetrates a short way into the metal. This so-called evanescent wave dies away rapidly beneath the surface; but if the metal is a very thin film it is possible for the evanescent wave to emerge, with diminished strength, on the far side. When it emerges, it resumes its behavior as an ordinary light wave. In effect, the light has traversed the thin film of metal. Thus, a very thin metallic film will be translucent. This ability of waves to jump across, or tunnel through, a thin reflecting barrier is quite general—it also happens, for example, with sound waves. In the case of alpha particle waves, it serves to cause a tiny "leakage" of the wave through the nuclear force barrier into the region outside. As we shall see, this implies that there is a small, but nonzero, probability of an individual alpha particle tunneling through the barrier and escaping. Given enough time, this is bound to occur.

But what about the mystery of why the same alpha particles cannot get back into the nucleus? The key to the answer lies in the words "given enough time." The degree of wave penetration of the barrier is usually so small that it may take an individual alpha particle billions of years to tunnel through. We notice such a slow phenomenon as the alpha decay of uranium only because even a small piece of uranium contains many trillions of nuclei, each with an alpha particle struggling to get out. The way the probabilities work out, if there is a billion-to-one chance of an alpha particle escaping from any one nucleus in a year, then if we watch *either* one nucleus for a billion years *or* a billion nuclei for a single year there is a good chance we will see one alpha decay. Watch 1,000 billion nuclei for a year, and we will expect to see about 1,000 decays; and so on. To actually observe the process go in reverse, however, one would need either to bombard the nuclei with trillions of alpha particles and hope to spot the odd one that penetrates the barrier, or to somehow bind an alpha particle tightly to the outside of the nuclear barrier, and then wait for several billion years.

An uncertain world

Weird as the tunnel effect is, it is perhaps even more astonishing to find that this effect is turned to practical use in everyday modern electronic devices—for example, those that incorporate components known as tunnel diodes. Perhaps the most spectacular demonstration of the wave nature of electrons is the phenomenon of superconductivity. When an electric current passes through an ordinary conductor such as copper, electrons migrate through the structure of the metal in a fairly haphazard way, often encountering irregularities and being scattered aside. This results in the familiar effect of electrical resistance. However, certain materials, when they are cooled close to the absolute zero of temperature (0 K, or about $-273°C$), suddenly lose all their resistance and become superconducting. A current can flow around a superconducting ring forever, without dissipating any energy.

The key to the remarkable property of perfect conduction lies with the wave nature of electrons. Each electron has its own electromagnetic field, which slightly distorts the crystal lattice of the material in which it is embedded; in turn, the distortion in the lattice of charged particles deforms their electromagnetic field, and thereby affects the other electrons. As a result of this, there is a very weak effective interaction between the electrons carrying the current through the crystal lattice. At ordinary temperatures, vibrations of the crystal lattice caused by heat overwhelm this tiny effect; but at very low temperatures the thermal vibrations are stilled, and the association between electrons comes to the fore. The association enables electrons to pair up with one another, and this pairing dramatically alters their properties. One effect is to enable large numbers of electron pairs to adopt the same wave configuration, effectively creating a gigantic electron superwave. Under the right circumstances, the electron superwave can go right around the ring of a macroscopic-sized superconductor, forming a circular wave that has settled into a fixed energy state from which it cannot be dislodged,[1] just like the stable orbit of an electron around the nucleus of an atom. Superconductors thus resemble, in some respects, macroscopic atoms. Like most quantum systems, they have been put to use in various practical ways, notably in making very powerful magnets for body scanners and other devices.

The wave properties of electrons are exploited in many other practical ways, too. The electron microscope, for exam-

1 Unless, that is, the temperature is increased.

ple, substitutes electrons for waves, because electron waves can have much shorter wavelengths than visible light, and therefore they can resolve much finer detail. Electron and neutron wave effects are used to investigate defects in metallic structures. And a beam of neutron waves directed at a target can be fine-tuned in frequency to resonate with the natural internal frequencies of the target nuclei; this seemingly esoteric trick makes it possible, among other things, to measure the temperature of the blade of a turbine in a jet engine while the engine is running.

The strangest thing about the dual wave-particle nature of the quantum world, however, is that it is not restricted to atomic and subatomic phenomena. In principle, even macroscopic objects such as people and planets have their individual quantum waves, determined by de Broglie's wave equation. The reason we never notice those waves (why, for example, people, do not "tunnel through" the chairs on which they sit, and fall to the floor) is contained in the formula itself—the length of the waves diminishes in proportion to the momentum. So the greater the mass of the object involved, the shorter the waves. Thus the wave of an electron in a domestic appliance is about one millionth of a centimeter long. A typical bacterium would have a wavelength less than the size of an atomic nucleus, and a pitched baseball has a wavelength of only 10^{-32} centimeters. Each of these objects can only tunnel through a barrier comparable in thickness to their respective wavelengths. When it comes to people and planets, the waves are so ridiculously short that for all practical purposes they can be ignored.

There are, however, deep issues of principle connected with the fact that matter waves exist even for macroscopic bodies, however short their wavelength might be, and scientists have wrestled with the problems for decades. It all goes back to a very basic question: what exactly *are* the quantum waves?

It is hard to see how something can be both a wave and a particle at the same time, and the discovery of the dual nature of both light and electrons caused a great deal of puzzlement at first. When physicists began to speak of wave-particle duality, they meant not that an electron was *both* wave *and* particle simultaneously, but that it could manifest *either* a wave or a particle aspect depending on circumstances.

Bohr extended the idea of wave-particle duality into something known as the principle of complementarity, which recognizes that seemingly incompatible physical qualities might be *complementary* rather than *contradictory*. Thus the wave and particle nature of electrons can be regarded as complementary aspects of a single reality, like the two sides of a coin. An electron can behave sometimes as a wave and sometimes as a particle, but never as both together, just as a tossed coin may fall either heads or tails up, but not both at once.

It is important to resist the temptation to regard electron waves as waves of some material substance, like sound waves or water waves. The correct interpretation, proposed by Max Born in the 1920s, is that the waves are a measure of *probability*. One talks of electron waves in the same sense as crime waves. To say that a city suburb is hit by a crime wave means that there is a greater likelihood that a burglary, say, will occur in a particular district. Similarly, the best place to look for an electron is where the electron wave is strongest—there is the greatest probability of finding an electron in that location. But even so, the electron *might* be somewhere else.

The fact that electron waves are waves of *probability* is a vital component of quantum mechanics and an important element in the quantum nature of reality. It implies that we cannot be *certain* what any given electron will do. Only the betting odds can be given. This fundamental limitation represents a breakdown of determinism in nature. It means that identical electrons in identical experiments may do different things. There is thus an intrinsic uncertainty in the subatomic world. This uncertainty is encapsulated in the uncertainty principle of Werner Heisenberg, which tells us that all observable quantities are subject to random fluctuations in their values, of a magnitude determined by Planck's constant. Einstein found the concept of quantum indeterminism so shocking that he dismissed it with the retort that "God does not play dice with the Universe!" and spent much of the rest of his life looking for the deterministic clockwork that he thought must lie hidden beneath the apparently haphazard world of quantum mechanics. That clockwork has not been found; it seems that God *does* play dice.

Bohr admonished those who would ask what an electron *really* is—wave or particle—by denouncing the question as meaningless. To observe an electron, one has to conduct some form of measurement on it, by carrying out an experiment ("tossing the coin"). Experiments designed to detect waves always measure the wave aspect of the electron; experiments designed to detect particles always measure the particle aspect. No experiment can ever measure both aspects simultaneously, and so we can never see a mixture of wave and particle.

A classic example is provided by a famous experiment first performed by Thomas Young in England in the early nineteenth century. Young carried out his experiment with light, but an exactly equivalent experiment has now been performed using electrons. In the original experiment, a point

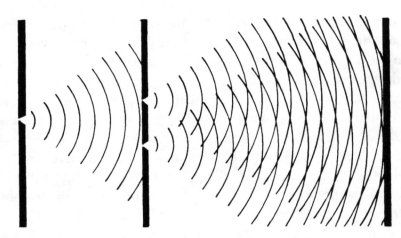

Figure 33. In Young's experiment, light from a point source (a pinhole in the first screen) passes through two nearby slits (in the second screen) to produce an image (on the third screen). The image shows alternate bright and dark bands, called interference fringes.

source of light illuminates two narrow adjacent slits in a screen, and the image of the light that passes through the slits is observed on a second screen (Figure 33). You might guess that the image would consist of two overlapping patches of light; in fact, it is made up of a series of bright and dark stripes, known as interference fringes.

The appearance of interference fringes in Young's experiment is a clear demonstration of the wave nature of light. Wave interference occurs in any wave system when two (or more) waves come together and overlap. Where the waves arrive in step they reinforce each other; where they are out of step they cancel each other. In Young's experiment the light wave from one slit intersects the light wave from the other slit to produce the bright and dark stripes, as the two waves alternately add together and cancel each other out. And it is important to appreciate that if either one of the slits is covered, the striped pattern disappears.

Paradoxical overtones emerge if one now regards the light as composed of particles—photons. It is possible to weaken the light source until only one photon at a time passes through the slit system, and to record the cumulative effect of many photons arriving one after the other at the second screen over a long period of time. Each photon arrives at the image screen and makes a spot on a photographic plate. In the equivalent electron experiment, single electrons are fired through a double-slit system, and the "image screen" is a sensitive surface like that of a television screen. The arrival of each electron makes a spot of light on the screen, and a video of the buildup of the spots of light shows how a pattern emerges as more and more electrons pass through the system.

Recall that one cannot know in advance, because of the inherent uncertainty of the system, precisely where any given photon or electron will end up. But the cumulative effect of many "throws of the quantum dice" will average out the distribution into a well-defined pattern. Moreover, this pattern shows the *same* series of interference bands as obtained with a strong source. The puzzle is this. Each particle, be it photon or electron, can clearly pass through one slit alone. And each particle, as the buildup of spots on the image screen indicates, behaves like a *particle* when it arrives, striking the screen in just one place. So how can an individual particle, which can pass through only *one* of the slits, "know" of the existence of the *other* slit and adjust its behavior accordingly? Could it be that a wave of *something* passes through the two slits, only to collapse into a particle when its position is "measured" by the screen? This is surely too conspiratorial, for the electrons or photons would have to know our intentions. And how does each individual particle "know" what the others will do so it can decide where it belongs in the interference pattern that builds up from the flow of thousands or millions of individual particles through the experiment? This is clear evidence for the holistic nature of quantum systems, with the behavior of individual particles being shaped into a pattern by something that cannot be explained in terms of the Newtonian reductionist paradigm.

Bohr expressed the situation clearly. Suppose we attempt to uncover the particle nature of photons by pinning down the location to the extent that we can tell through which slit each one passes. Then the result of this scrutiny is to smudge out the very interference pattern that is the hallmark of the wave aspect. Thus if we set up the experiment so that a counter sits at each of the two slits to record the passage of each photon through one slit or the other, the effect of making these observations is to introduce an additional uncertainty (via Heisenberg's principle) into the behavior of the particles. The magnitude of this uncertainty is just such that it smears out the interference pattern, leaving instead two superimposed smudges of light, just as you would expect for particles going through either of the slits without interference. So in exposing the particle aspect of the wave-particle duality, we destroy the wave aspect. We must therefore contend with two different experiments, one revealing the wave aspect and the other the particle aspect. The results of the experiment depend on the nature of the whole experimental setup, apparatus plus light (or electrons), and not just on the nature of light itself. These ideas may seem to defy common sense—but remember, our common sense is based on experience of things much bigger than photons or atoms, and there is no reason why it should be a good guide to what goes on at the atomic level.

Creating reality

As if this were not bewildering enough, a further twist was added by John Wheeler, of the University of Texas at Austin. He pointed out that the holistic nature of reality extends not just through space but through time as well. Wheeler showed how a decision as to which of the two complementary aspects of reality—wave or particle—shall be revealed by the two different double-slit experiments can be left until *after* the photon (or electron) has already passed through the double-slit system. It is possible to "look back" from the position of the image screen to find out which slit any given particle has come through. Alternatively, one could choose not to look, and leave the interference pattern to develop as usual. The decision of the experimenter about whether or not to look back *at the time the particles arrive at the screen* determines whether or not the light *was* behaving in the manner of particles or waves at an earlier moment, at the time when it passed through the slit system in the first screen.

Wheeler called his arrangement the "delayed-choice" experiment. A practical version was carried out at the University of Maryland by Caroll Alley, and it completely confirmed Wheeler's ideas. The apparatus involved a system of laser beams on a laboratory bench, and although in this case the "choice" was delayed by only a few billionths of a second, an important principle was established as fact. Wheeler has extrapolated that principle to the extreme case where nature provides a sort of cosmic two-slit system. The gravitation of a black hole, or a galaxy, or even a cosmic string can bend and focus light like a lens, as we described in Chapter 6. Figure 34 shows how a distant astronomical object can provide for

Figure 34. A massive object such as a galaxy, or even a black hole, can act as a giant lens. Light from a distant source is bent by the gravitational space warp surrounding the object. This effect, equivalent on a grand scale to the bending of starbeams by the Sun (Figure 16), can produce multiple images of a distant source, like those produced by the effect of a cosmic string (Figure 29).

two light paths through space to converge. One might imagine a remote source of light, such as a quasar, emitting photons that pass around some intervening object and are focused at the Earth. The two paths around the object then play the role of the two slits. An experimenter on Earth could in principle bring the two light beams together in an interference experiment. If the delayed-choice facility were now deployed, the decision of the experimenter to expose either the wave or particle nature of the quasar light would affect the nature of that light—not just a few billionths of a second in the past, but *several billion years* ago! In other words, the quantum nature of reality involves nonlocal effects that could in principle reach right across the Universe and stretch back eons in time.

It is important to realize, however, that the delayed-choice experiment does not provide the capability for sending *information* back into the past. You could not, for example, use the experiment to signal another experimenter located near the distant quasar several billion years ago; any attempt by that distant experimenter to investigate the condition of the quasar light as it passed by, thereby attempting to read the signals from the future, would inevitably disturb the quantum state and destroy the very signal that the Earth-based experimenter is attempting to send back from the future. Nevertheless, the delayed-choice experiment illustrates graphically that the quantum world possesses a kind of holism that transcends time, as well as space, almost as if the particle-waves seem to know ahead of time what decision the observer will make.

Probably the most unsettling aspect of these studies is the way that the observer seems to play a central role in fixing the nature of reality at the quantum level. This has long worried both physicists and philosophers. In the prequantum era of physics, everyone assumed that the world "out there" existed in a well-defined state quite irrespective of whether, or how, it was observed. Admittedly the act of observation would intrude into that reality, for we cannot observe anything without interacting with it physically to some extent; yet it was always supposed that the interaction was purely incidental and could either be made arbitrarily small (at least in principle) or else be performed in a controlled way and so be taken precisely into account. But quantum physics presents a picture of reality in which observer and observed are inextricably interwoven in an intimate way. The effect of observation is absolutely fundamental to the reality that is revealed, and cannot be either reduced or simply compensated for.

If, then, the act of observation is such a key element in creating the quantum reality, we are led to ask what actually happens when an observation of an electron or a photon is made. As we have mentioned, the wave nature of macroscopic objects is usually insignificant in everyday life. However, it seems that during a quantum measurement the wave properties of the measuring apparatus, and even of the observer, cannot be ignored.

The role of the observer is highlighted by what is known

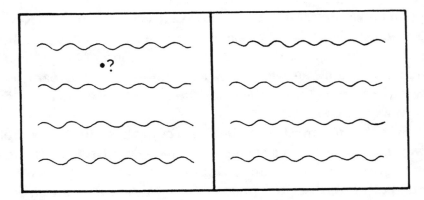

Figure 35. An electron is placed in a box, and then a partition is inserted. According to quantum mechanics, the wave associated with the electron will occupy both compartments of the box, reflecting the possibility that the electron might be in either compartment. "Common sense," however, tells us that the electron, being a particle, must be in either one compartment or the other.

as the measurement paradox. Suppose, for the sake of argument, that the wave corresponding to an electron is confined to a box and the particle is equally likely to be found anywhere inside the box. Then imagine that a partition is slid into the box, dividing it into two equal halves (Figure 35). According to the quantum rules, the wave is still present in both halves of the box, reflecting the fact that when we look for the electron we are equally likely to find it on either side of the partition. Common sense, however, would dictate that the electron can be in only either one half of the box or the other. Suppose, now, that someone looks inside the box and finds the electron in one particular half. Clearly the probability wave must abruptly disappear from the other half of the box, because it is now known with certainty to be empty.

The oddity of this abrupt resculpturing of the wave—often called "the collapse of the wave function"—is that it seems to depend upon the activities of the observer. If nobody looks, then the wave never collapses. So the behavior of a particle such as an electron appears to vary according to whether it is being watched or not. This is deeply troubling to physicists, but may not seem of any great concern to other people—who else really cares what an electron is doing when we are not looking at it? But the issue goes beyond electrons. If macroscopic objects also have associated waves, then in principle the independent reality of *everything* seems to go into the quantum melting pot.

Many physicists feel very uneasy about large systems having wave properties that play a part in the outcome of experiments. One reason for their concern is that it is possible to envisage arranging for two waveforms which represent very different macroscopic states to overlap and interfere with one another. The most famous example of this was dreamed up by Schrödinger. It consists of a cat incarcerated[2] in a box containing a flask of cyanide and a hammer poised above the glass (Figure 36). A small source of radioactivity is arranged so that if, after a certain period of time, an alpha particle is emitted, this is detected by a Geiger counter and triggers the fall of the hammer, which breaks the flask and kills the cat. The scenario provides a memorable demonstration of the paradoxical nature of quantum reality.

One can imagine a situation in which, after the specified time, the alpha particle's wave lies partly within the nucleus and has partly tunneled out. This might correspond, for example, to equal probability that the alpha particle had, or had not, been ejected by the nucleus. Now the rest of the stuff in the box—Geiger counter, hammer, poison and the cat itself—can also be treated as a quantum wave. One can

2 Hypothetically, we hasten to add!

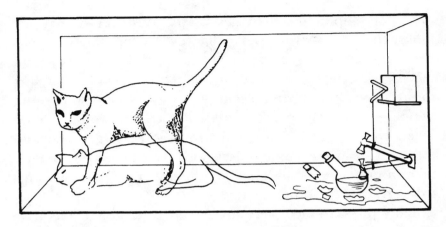

Figure 36. A schematic illustration of Schrödinger's cat experiment, showing the ghostly superposition of live and dead cats. (Cat lovers please note—this is a thought experiment only!)

therefore envisage two possibilities. In one case the atom decays, the hammer falls, and the cat is dead. In the other case, which has equal probability, none of this happens and the cat remains alive. The quantum wave must incorporate all possibilities, so the correct quantum description of the total contents of the box must consist of two overlapping and interfering waveforms, one corresponding to a live cat, the other to a dead cat. In this ghostly hybrid state, the cat cannot be regarded as definitely either dead or alive, but in some strange way both. Does this mean we can perform the experiment and create a live-dead cat? No! If the experimenter opens the box, the cat will be found to be *either* alive or dead. It is as if nature suspends judgment on the fate of the poor creature until somebody peeks. But this raises the obvious question: *what is going on inside the box when nobody is looking?*

It is clear from scenarios such as this that the wave properties of matter applied to macroscopic objects—and especially to conscious observers—raise very deep issues about the nature of reality and the relationship between the observer and the physical world. The cat scenario is deliberately contrived to tease out the paradoxical nature of quantum weirdness in a dramatic way, but the same essential phenomenon occurs every time an alpha particle is emitted by a nucleus, and is busily at work in the radioactive paint on the hands of your luminous clock.

There is still no general agreement on how to resolve paradoxes like that involving Schrödinger's cat. Some physicists believe that quantum mechanics will fail for systems as large and complex as cats. Another opinion is that quantum physics can tell us nothing about individual alpha particles or cats, but only about the statistics of collections of identical systems, so that we can say that if we were to perform the same experiment with a thousand cats in identical boxes then a certain fraction of the cats (as determined by the quantum rules) will be found alive and the rest dead. But that simply dodges the question of what happens to any individual cat.

Perhaps the most dramatic attempt to make sense of such quantum superpositions is the so-called many-universes (or alternative-histories) theory. In the context of the cat experiment, this states that the entire Universe splits into two coexisting, or parallel, realities, one with a live cat and the other with a dead cat. Although it may seem like science fiction, the many-universes theory is entirely consistent with the rules of quantum mechanics and is supported by several leading theoretical physicists. We shall take a closer look at this theory shortly.

The theory of parallel worlds developed, as we have seen, out of the fundamental paradox concerning the nature of reality as it pertains to the world within the atom. Because of the wave-particle duality of entities such as electrons, it is impossible to attribute to them precisely certain properties, such as possessing a well-defined path through space, that we are used to thinking of in connection with macroscopic objects like a bullet or a planet in its orbit. Thus, when an electron goes from A to B, its trajectory is fuzzed out by quantum uncertainty, as described by Heisenberg's uncertainty principle. In one form, this principle states that you cannot know, at any instant, *both* the position and the momentum of a quantum particle. Indeed, it goes deeper—it says that a quantum particle *does not possess* both a definite momentum and a definite position simultaneously. If you try to measure accurately the position, you lose information about the momentum, and vice versa. There is an irreducible trade-off between these two qualities. Either can be known as accurately as you like, but only at the expense of the other.

We have encountered the uncertainty principle in our discussion of quantum chaos, the nature of the vacuum and the origin of time. This is the same uncertainty that also affects energy and time, and tells us that virtual particles can pop briefly out of nothing at all, and vanish again. Such quantum uncertainty is not merely a result of human clumsiness. It is an *intrinsic* quality of nature. However accurate and powerful our instruments may be, we cannot beat the inherent fuzziness of quantum uncertainty.

The trade-off between position and momentum is another example of quantum complementarity at work. It turns out to bear a close relation to the wave-particle complementarity. The wave associated with an electron is, by its very nature, a spread-out thing, and does not have a definite position, although it does encode information about the electron's momentum. By contrast, the particle associated with an electron is, by its very nature, something with a well-defined position; but a wave collapsed to a point carries no information about the momentum of the electron. Measure the position of an electron, and you do not know *(nor does the electron know)* how it is moving; measure the momentum of an electron, and neither you nor the electron know where it is located.

Einstein's dilemma

In the early days of the quantum theory these strange results divided physicists into two camps. There were those, led by Niels Bohr, who fully accepted the implications of the theory, and insisted that the microworld is inherently indeterministic. And there were those, most notably Einstein, who maintained that quantum mechanics could not be regarded as a satisfactory theory if it made such nonsensical claims. As we have mentioned, Einstein hoped that behind the weird quantum world lay a hidden reality of concrete objects and forces moving in accordance with the more traditional notions of cause and effect. Einstein supposed that the fuzziness of quantum systems is somehow a result of observational inadequacy. Our instruments are simply not elaborate enough, he believed, to reveal the intricate details of the variables that determine the seemingly erratic behavior of subatomic particles. Bohr's view was that there *are* no causes of this chaos, that the old Newtonian view of a clockwork Universe unfolding according to a predetermined pattern is forever discredited. Rather than rigid rules of cause and effect, claimed Bohr, matter is subject to the laws of chance. The processes of nature are not so much a game of pool as a game of roulette.

Much of the early debate about quantum reality focused on certain "thought experiments," like that involving Schrödinger's cat. Einstein hoped to devise a scenario in which the rules of quantum mechanics would lead to an inconsistency or absurdity. He used to contrive imaginary situations that would seem to threaten Bohr's position, only to find that Bohr devised an escape route. Eventually, Einstein gave up trying to demonstrate that quantum mechanics was inconsistent, and concentrated instead on trying to show that it was incomplete. In other words, Einstein might reluctantly have conceded that quantum mechanics was the truth, but he would never accept that it was the *whole* truth.

The claim of incompleteness turned on the nature of quantum uncertainty. Einstein wanted to believe that, say, an electron *really did have* both a well-defined position and a well-defined momentum at the same time, even though in typical practical experiments knowledge of one aspect might frustrate attempts to know the other. He tried to conceive of a way of demonstrating that "elements of reality" could be attached simultaneously to *both* complementary qualities. His best attempt, formulated with colleagues Nathan Rosen and Boris Podolsky, sought to obtain information about both the position and the momentum of a particle by using an accomplice particle. As a second particle bounced off the particle we are interested in, the accomplice would retain some information about the position and motion of the first particle, much as the rebounding pool balls from a break carry information about the speed and direction of the cue ball that struck them. Observe one particle out of a pair involved in a collision, and you can infer something about the other by reconstructing the collision mathematically.

Suppose, Einstein reasoned, that there are two particles, A and B, which collide and separate to a great distance. Now we are free to measure either the position or the momentum of B. If we measure the former we can infer the position of

A, from the laws that govern collisions. But we could equally well measure the momentum of B and use it to infer the momentum of A. Einstein suggested that although a measurement of B's position might fuzz out the momentum of that particle (or vice versa), the act of a measurement on B could not immediately affect particle A, which might be a long way away by the time the measurement is made. At the very least, no physical influence from the measurement of B could reach A in less time than it would take light to travel from B to A—the ultimate speed limit of Einstein's own theory of relativity. Certainly, it seemed to Einstein that at the instant of the measurement of B, the state of particle A must remain undisturbed.

This seemed to settle the issue, for if the experimenter chose to measure *either* the position or momentum of B, and hence infer *either* the position or momentum of A, in either case without disturbing A, then surely A must *already possess* both "elements of reality" at the time of measurement. Indeed, one could envisage measuring the momentum of A by this proxy technique (that is, by measuring the momentum of B and inferring that of A) and *at the same instant* conducting a position measurement directly on A, thereby yielding precise values for both quantities at the same time. So, Einstein reasoned, it is possible in principle to know the position and the momentum of particle A at the same time. It seemed to Einstein that the only way to retain quantum uncertainty across the gap between the particles would be if they were connected by what he called some "spooky action at a distance," operating faster than light and therefore transcending the restraints of his own theory of relativity.

Although Bohr provided a defense of his position in the face of this formidable challenge, the case rested until the 1960s as a pure thought experiment. Then John Bell of

CERN extended the Einstein-Podolsky-Rosen experiment to a wider class of two-particle processes, producing general rules that all such systems must obey if they are to comply with Einstein's "common sense" picture of reality. Bell found that these rules incorporate a mathematical restriction now known as Bell's inequality. For the first time, it became possible to consider an actual laboratory test of these ideas. If the experiments showed that Bell's inequality is obeyed, Einstein would be proved right; but if the inequality was violated, Einstein would be proved wrong. Following Bell's work, a series of careful experiments has been performed, culminating in an accurate test of Bell's inequality by Alain Aspect, of the University of Paris, in 1982. Aspect's experiment consisted of performing simultaneous measurements on pairs of oppositely directed photons that were emitted in a single event from the same atom, and so possessed correlated properties. The results? Einstein was wrong. This conclusion has since been confirmed by repeated experiments. What does it mean?

Assuming one rules out faster-than-light signaling, it implies that once two particles have interacted with one another they remain linked in some way, effectively parts of the same indivisible system. This property of "nonlocality" has sweeping implications. We can think of the Universe as a vast network of interacting particles, and each linkage binds the participating particles into a single quantum system. In some sense the entire Universe can be regarded as a single quantum system. Although in practice the complexity of the cosmos is too great for us to notice this subtle connectivity except in special experiments like those devised by Aspect, nevertheless there is a strong holistic flavor to the quantum description of the Universe.

The Aspect experiment essentially lays to rest Einstein's

hope that quantum uncertainty and indeterminism can be traced back to a substratum of hidden forces at work. We have to accept that there is an intrinsic, irreducible uncertainty in nature. An electron or other quantum particle generally *does not have* a well-defined position or motion unless an actual measurement of position or motion is made. The act of measurement causes the fuzziness to give way to a sharp and definite result. It is this combination of uncertainty and of the collapse of the quantum wave when an observation is made that leads to the paradox of the cat in the box. But so far we have looked only at a very simple version of this puzzle. What happens when we try to apply what we have learned from it to the Universe at large?

The very concept of a superposition of live-cat and dead-cat states waiting to be resolved when someone looks in the box seems absurd, because presumably the cat itself knows whether it is dead or alive. Does this knowledge not constitute an observation leading to a collapse of the quantum wave into a definite state one way or the other? Surely it is not necessary for all quantum observations to be conducted by human beings before they can be regarded as producing a definite state of reality? But if a cat can do the job, what about an ant? Or a bacterium? Or can we dispense with a living component in the experiment altogether, and leave it all up to a computer, or even a camera?

As far as the world outside the box is concerned, however, we can regard the whole laboratory as just a bigger box. If the experimenter has looked inside the box and determined the fate of the cat, a colleague working in the lab next door may not know this. Does the quantum wave of the whole lab collapse only when the colleague opens the door and asks how the cat is getting on? This clearly leads us into an infinite regress. Each quantum system can be collapsed into a definite state on being measured by another system outside itself, but then the larger system goes into an indeterminate state and must be collapsed into reality by yet another system outside of it, and so on.

Various resolutions have been proposed to break out of this deadlock. According to one highly speculative point of view, it is necessary to invoke the concept of mind at some stage, and to argue that the chain of regress (similar to the regress in Dunne's theory of serial time) ends when the result of the measurement enters somebody's consciousness. This endows the world with an extremely subjective element, for it obliges us to suppose that, in the absence of observation, the external world does not exist in a well-defined sense. It is as though through our observations we actually create, rather than explore, the external world.

Many physicists are content to ignore the infinite regress, on the basis that however large their laboratory may be, there is still a lot of the Universe outside it to cause the collapse of the laboratory contents into concrete reality. Cosmologists, however, do not have this option. Their laboratory is the Universe itself, and there is nothing outside the Universe which can observe it.

Multiple realities

This is where the many-worlds interpretation seems to force itself on us. In terms of serious physics, as opposed to the pages of science fiction, the idea dates from 1957, with the work of an American, Hugh Everett. It has since been refined by others. As we remarked earlier, the many-universes theory resolves the cat paradox by supposing that the Universe divides into two copies, and both then coexist in parallel with each other. There is thus no impediment to applying quantum mechanics to the entire Universe, if we are prepared to entertain the rather fantastic notion that the whole Universe is continually splitting into countless copies, each in a slightly different state, one for every possible outcome of every possible quantum interaction. The Everett theory suggests a sort of multiple reality, in which an infinite number of entire universes coexist. Bizarre though this may seem, the actual mathematical formalism involved is identical with standard quantum mechanics. The novelty of the theory concerns only the *interpretation* of the quantities that appear in the equations.

An obvious objection to the many-universes theory is that we experience only one reality, one Universe. Where are all the others? To understand the answer to this question we need to take on board the concept of spacetime discussed earlier in this book. When the Universe divides into many copies, the splitting creates many duplicates not only of material objects but of space and time as well. That is, each "new" universe comes into being with its own space and time. The other worlds are not, therefore, "out there" in any everyday sense of the term. They cannot be reached through our own space and time. Instead, they are complete spacetimes in themselves. When we ask "where" something is, we normally assume that the thing is located at a certain distance and direction from wherever we are. But the Everett universes are not "in" our Universe at all. They do not lie at any particular distance or in any particular direction from us.

It may be difficult to picture this. But the fact that we cannot visualize many different spacetimes does not, of course, logically preclude their existence. We are still able to describe the other universes mathematically. Nevertheless, some sort of imagery is helpful. One possibility is to regard the many universes as "stacked up" like the pages of a book laid flat on a table. In this collection of two-dimensional sheets, each page represents an entire universe—that is, spacetime plus matter. The form of each universe is very slightly different from its neighbors according to the different quantum alternatives realized therein. As we move farther down the stack, away from the page chosen as our reference point, the differences accumulate.

Sometimes the many-universes model is represented by the branches of a tree. The "trunk" symbolizes a particular universe which we use as our reference point; this then branches and rebranches into all its quantum alternatives. We can imagine a horizontal slice through these multiple branches at some moment, intersecting a whole collection of slightly different universes which have "grown" out of the original. And in general, the trunk itself will be just one branch from a more elaborate tree which extends to infinity.

When people first learn about the many-worlds theory they often object that it cannot be true because we do not notice any such splitting taking place. However, an important feature about the theory is that human observers do not play an especially distinguished role; they too are split like everything else. In the cat example, where the Universe divides into two copies, one containing a dead cat and the other a live cat, each "new" universe contains one copy of the experimenter as well. Each copy of the experimenter looks into the box to see the cat's fate. One experimenter, in one universe, sees a live cat; the other experimenter, in the other world, sees a dead cat. Each experimenter mistakenly believes that his or her universe is unique, and that the reality perceived on opening the box (dead cat or live cat) is the only reality.

Taking the many-worlds theory to its logical conclusion, we are led to suppose that countless times every second each human being is split into duplicate copies, each copy inhabiting a slightly different universe. Necessarily, each copy will only perceive *one* universe, and be aware of only *one* self.

There is a somewhat different approach to the many-universes idea, which does not involve any actual splitting. In this version there are always the same number (in fact, an infinite number) of parallel realities, but at any given time many copies will be precisely identical. For example, in the cat experiment, one could consider two universes coexisting before the experiment, but being completely indistinguishable. At the point where the experiment with the cat is performed, these two worlds differentiate into one in which the cat remains alive and the other in which the cat dies. From this perspective, as you read these words there are many identical copies of you inhabiting strictly identical universes (and yet other copies living in slightly different universes). In the future, however, some of these copies will cease to be identical as their respective universes evolve differently according to their different quantum choices. At a gross level, we can visualize this by imagining that out of the infinite number of versions of "you" reading these words, all with identical pasts, some will carry on reading the book, others will set it aside to go off temporarily to make a cup of coffee, still others will notice that the Sun is shining and decide to abandon reading for the day altogether, and so on.[3] Slightly more subtly, an infinite number of you will, as a result of a minor quantum fluctuation in the hardware of the typesetting computer used by our printers, notice that, say, the fourth word of the next paragraph is misprinted, while the rest of you will not, because that quantum fluctuation will not have happened in your universe.

The question naturally arrises as to whether it is possible in any way to travel into, or at least to communicate with,

---

3 It would be good to have the infinite sales of our book across the many worlds—but the catch, of course, is that there are an infinite number of us writing the book, and we all have to share the income!

these other worlds. The answer is that, in the normal course of events, it is not. We cannot, alas, invoke parallel realities to explain ghosts, or ESP, or UFOs. Indeed, the whole point of the Everett theory is that the different branches of the universe(s) are physically disconnected, alternative realities. This is necessary, in order to resolve the paradox of quantum measurement and to avoid feeling ourselves split.

But, as our examples two paragraphs back make clear, a measurement as we normally understand it takes place when we become aware of some macroscopic change, such as a click in a Geiger counter or the position of a pointer on a meter (or the state of health of a cat). Our brains register these events in a sharply defined way because both the apparatus and presumably our brain cells are macroscopic entities for which quantum effects are negligible. It is possible, though, to conceive of a conscious individual whose sensory perception and memory operate at the quantum level. In fact, computer scientists are currently researching the idea of using switching devices built at a molecular scale in an effort to achieve still greater miniaturization than the present generation of computers. The British physicist David Deutsch has proposed a remarkable experiment based on this prospect, which actually seems to permit some sort of rudimentary contact to be established between parallel worlds.

In the Deutsch experiment a quantum brain (be it natural or artificial) is required to carry out a typical double-choice quantum experiment. For example, it could observe whether an electron bounces off to the left or to the right from a fixed target. According to the many-worlds theory there is a universe with a left-moving electron and another with a right-moving electron.

Now, as observed by us, when two universes split or become differentiated, they do so irreversibly. We cannot, at a macroscopic level, perceive subsequent developments in such a way that the universes remerge, or become identical again. An event such as the death of a cat is clearly an irreversible occurrence. But at the atomic level it is perfectly conceivable for changes to be reversed. One can easily devise an atomic experiment in which a quantum particle undergoes some double-choice experience, but in which the state of the particle is afterward returned to its initial form.

In short, at the atomic level worlds *can* be split and merged again by careful manipulation. These temporary hybrid states are not seen by us as separate alternatives, because as soon as we attempt to observe them we introduce irreversible macroscopic influences which split the worlds permanently. The quantum brain envisaged by Deutsch, however, can observe things without causing this irreversible split. A quantum brain could register the hybrid reality without preventing the remerging of the temporarily split worlds. During the temporary split, the brain would indeed divide into two copies, but these would merge again after the experiment. Each copy would carry a different memory of the behavior of the electron that was being observed. The remerged brain would therefore be endowed with a double memory. It could tell us what events had been like in *both* possible worlds. In this simplified way, we really could obtain some information about more than one reality.

The proposed Deutsch experiment depends upon the existence of a quantum-level intelligence, and although such ideas are taken seriously by some artificial intelligence experts, all are agreed that it will be a long time before we can expect to build such a thing. Meanwhile, it is interesting to ask whether there is any indirect evidence for the existence of a multiple reality.

Cosmic coincidences

Over the past few years a growing number of physicists and astronomers have become impressed by the fact that the Universe we perceive seems possessed of a remarkable range of apparent accidents of fortune. A few examples will suffice to give an idea of what is involved (further discussion can be found in our books *The Accidental Universe* and *Cosmic Coincidences*, detailed in the Bibliography.)

One of the most dramatic coincidences concerns the stability of atomic nuclei. Recall our discussion of alpha decay, with which we began our closer look at quantum weirdness. Atomic nuclei are held together, as we have seen, by a powerful nuclear force. The stability of the nucleus involves a competition concerning the strong nuclear force, the electromagnetic force and quantum effects associated with tunneling. There is a fairly narrow range of possible nuclear structures in which these competing influences balance in a stable fashion.

To take a specific example, due to Freeman Dyson, if the strong force were just a few percent stronger it would enable two protons to combine in a stable form, overcoming the mutual repulsion of their individual positive charges even without the buffering presence of a neutron or two. If such a di-proton were to form, one of the protons would soon decay into a neutron, converting the di-proton into a deuteron—a nucleus of deuterium. Deuterium is an efficient nuclear fuel, so such an eventuality would short-circuit the nuclear processes that take place in the core of the Sun and other stars, and lead to the wholesale explosive consumption of all the nuclear fuel in the Universe. Indeed, this would have taken place way back in the big bang, effectively depriving the Universe of free protons, and hence of the element hydrogen, since the nucleus of a hydrogen atom is a single proton. Without hydrogen (which actually constitutes the majority of the visible cosmic material) there would be no stable stars like the Sun and no water in the Universe. Life as we know it could not arise under those circumstances.

Equally dramatic consequences would ensue if the nuclear force were slightly weaker relative to the electric force, for then the element deuterium (whose nuclei consist of a single proton and a single neutron in combination) could not exist. Deuterium plays a vital role in the chain of nuclear reactions that keeps the Sun and stars burning. And similar delicate balances apply to other forces of nature.

Astrophysicist Brandon Carter has shown how the structure of stars is very delicately dependent on the exact ratio of gravitational to electromagnetic forces. Our Sun is a middling sized, yellow star, and the conditions that make life on Earth possible are closely dependent on the Sun's basic nature. If these forces were very slightly different in their relative strengths, however, all stars would be either blue giants or white dwarfs, depending on which way the balance of forces was tilted. Stars like our Sun, which seem to be ideal at providing conditions suitable for the emergence of life, would not exist.

These apparent "coincidences," and many more like them, have convinced some scientists that the structure of the Universe we perceive is remarkably sensitive to even the most minute changes in the fundamental parameters of nature. It is as though the elaborate order of the cosmos were a result of highly delicate fine-tuning. In particular, the existence of life, and hence intelligent observers, is especially sensitive to the high-precision "adjustment" of our physical circumstances.

For some people, the exceedingly fortuitous arrangement of the physical world, which permits the very special conditions necessary to human observers' existence, confirms their belief in a creative Designer. Others, however, point to the many-universes theory as a natural explanation for cosmic coincidences. If an infinite array of universes really does exist, each of which realizes a slightly different cosmic possibility, then any universe, however remarkable or improbable, is bound to occur somewhere in the array. It is no surprise then that the Universe (or universes) that we perceive is of this remarkable sort, for only in a cosmos in which the conditions necessary for life to form have occurred (by chance) will observers arise to ponder over the meaning of it all.

If these ideas are correct, they imply that the overwhelming majority of other universes are inhospitable, and go unobserved. Only in an exceedingly narrow range of possible worlds—a few pages out of the infinite cosmic book—will the many accidents necessary for life to form fortuitously combine, and so only an infinitesimal fraction of the total stack of universes is actually cognizable.

This type of reasoning, known as the anthropic principle, was considered briefly in Chapter 2 in connection with the laws of physics in general. It can offer only circumstantial evidence for the existence of parallel universes, but many scientists find it a preferable hypothesis to the belief in a supernatural design. Until we can build a quantum superbrain, the cosmic coincidences offer the best argument we have that, somewhere that is not anywhere, myriad other writers identical with us are writing books identical with this one, to be read by myriad identical readers, each following their parallel lives toward slightly different destinies—and now wondering about the existence of all their duplicates.

Further speculation along these lines is fruitless, until the advent of quantum intelligence. Meanwhile, equipped with a deeper understanding of quantum processes (and quantum weirdness) we can probe more deeply into the modern understanding of space and time.

8 The Cosmic Network

The matter myth is built on the fiction that the physical Universe consists of nothing but a collection of inert particles pulling and pushing each other like cogs in a deterministic machine. We have seen how in their various ways the different branches of the new physics have outdated this idea. Quantum physics especially pulls the rug out from under this simple mechanistic image. We discussed how quantum non-locality forbids one from considering even widely separated particles as independent entities. When quantum mechanics is extended to encompass the field concept—a branch of physics known as quantum field theory—it brings with it a wonderland of nebulous activity, such as virtual particles and the ferment of the vacuum. Even the apparent solidity of ordinary matter melts away into a frolic of insubstantial patterns of energy.

Quantum field theory creates an image of a universe criss-crossed by a network of interactions that weave the cosmos into a unity. As we have explained, physicists recognize the existence of four fundamental forces: electromagnetism, gravitation and the weak and strong nuclear forces. Three of these forces can be accurately described in terms of quantum field theory as part of the cosmic network. But gravitation has stubbornly resisted the efforts of theorists to cast it in this mold. This is widely regarded as a very serious shortcoming in our description of nature. As we have seen, the general theory of relativity closely links gravitation with the geometrical structure of spacetime, and forms one of the twin pillars of twentieth-century physics. Quantum mechanics constitutes the other pillar. Yet the fact is that the marriage between these two great theories remains unconsummated.

It is not possible simply to shrug this difficulty aside, because the very consistency of quantum mechanics requires that *all* of nature be subject to quantum rules. If it were not, we could devise gravitational experiments that would enable us to violate, say, Heisenberg's uncertainty principle. In recent years, however, physicists have become increasingly excited by the prospect of recasting gravitation in an entirely new guise, making it possible not only to find a consistent quantum description, but to marry the force of gravitation with the other three forces of nature to produce a unified superforce, thus providing a truly integrated cosmic network.

Photons light the way

In order to shed some light on the difficulties involved in constructing a quantum theory of gravity, it helps first to review the simpler case of electromagnetism, the archetypal quantum field theory. A charged particle, such as an electron, which is the source of an electromagnetic field, can be envisaged as a point of matter at the center of a field of unseen electromagnetic energy, surrounding it like a halo extending into space. When another electron approaches close to the first electron, it senses the field and experiences a repulsive force. It is as though the field of one electron sends out a message: "I'm here, so move."

The message travels through the field in the form of a disturbance, which exerts a mechanical effect both on the receiving particle (action) and on the transmitting particle (reaction). In this way, electrically charged particles act on one another across empty space. And, of course, in the classical picture of this process at work the messages linking all charged particles in a network of action and reaction are car-

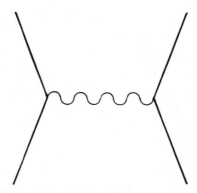

Figure 37. Two electrons interact by exchanging a virtual photon. The photon (wavy line) acts as a "messenger" conveying the electromagnetic force between the two electrons. The net result is to cause the two electrons to scatter off each other.

ried by ripples in the electromagnetic field, that is, by electromagnetic waves.

Quantum theory retains the basic idea of a field, but the details are radically altered. Electromagnetic disturbances, as we have seen, can be emitted or absorbed only in discrete packets, or quanta, known as photons, so we must envisage the disturbance in the electromagnetic field which conveys the interaction as involving the exchange of photons. These photons, in effect, carry the messages between electrons and other charged particles. Instead of envisaging the field of one electron continuously disturbing the path of another, we must picture instead the first electron emitting a photon which is then absorbed by the other (Figure 37). It is rather like firing cannonballs across space; the first electron recoils in response, while the second is deflected by the impact. The disturbance therefore takes place abruptly. An observer would see the end result as the scattering of one electron away from the other, and infer that their electric charges were causing a repulsion.

Although mathematically the description of this scattering process involves abrupt changes, these cannot actually be discerned in an experiment, nor can the passage of the photon be witnessed directly. This is because of the essential quantum fuzziness of subatomic systems required by the Heisenberg uncertainty principle. Electrons cannot be ascribed well-defined paths in space, and even the time order in which the photon is emitted and absorbed is imprecise. The messenger photons thus have a sort of ghostly fleeting existence. To distinguish them from the permanent sort that we experience directly through the sensation of sight, the messenger photons are referred to as "virtual." We encountered virtual photons and other virtual particles in Chapter 5, where their influence on the nature of the vacuum was discussed; they also play other roles in the quantum world.

Though we have described the process of electron scattering in terms of the exchange of a single photon between two charged particles, there is also the possibility that two, or more, photons will be exchanged (Figure 38). Two-photon exchanges turn out to have a weaker influence on the overall physical process than single-photon exchanges; three-photon exchanges are weaker still; and so on.

Although the individual details of the photon exchange can never be observed, the full mathematical treatment of these ideas does give clear predictions about things that *can* be observed, such as the average angle of scatter when two electron beams collide. In this respect, the virtual photon description of the electromagnetic force has been an astonishing success. The full details were worked out in the late 1940s, and given the name quantum electrodynamics, or QED for short. The theory permits calculations that predict

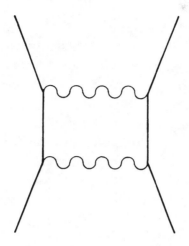

Figure 38. There is a small probability that two electrons will interact by exchanging two or more photons. Such processes lead to small corrections in calculations of the scattering efficiency of electrons.

some truly minute and subtle effects, such as a tiny shift in the energy levels of electrons in atoms caused by the presence of messenger photons. In some of these effects, the influence of multiple-photon exchange has to be taken into account. Ingenious experiments have confirmed these effects to astonishing accuracy—the latest versions of the experiments are precise to one part in 10 billion, and find perfect agreement with the theory. Such stunning success has prompted the proclamation that QED is science's most successful quantitative theory.

A network of messengers

What we normally think of as empty space is in fact continually crisscrossed by an incessant traffic of messenger particles such as virtual photons. The rate at which messenger traffic proceeds depends upon the strength of the force concerned.

Strong forces are the source of very frequent traffic, weak forces less so. If it were not for the unending network of messenger traffic, particles of matter would be completely oblivious of one another; there would be no interactions at all. Every particle would simply fly off along its own independent path through space, never deviating, and to all intents and purposes it would be alone in the Universe. Composite objects could not exist, because no force would be available to bind them together.

The basic idea behind QED—the exchange of messenger particles—has been successfully extended to the quantum description of the two nuclear forces. These forces each have their associated fields, which can be described in terms of messenger particles, analogous to photons. In the case of the weak force, the particles involved, although long predicted by theory, were discovered only in 1983, and they are cryptically known as W and Z particles. The case of the strong nuclear force is a little more complicated. The nuclear particles (protons and neutrons) are now known to be composite objects, each made up of three smaller units called quarks. The quarks are bound together by a very strong force which deploys no less than eight messenger particles, dubbed gluons. The force that binds neutrons and protons within nuclei is a weaker vestige of this powerful inter-quark gluon force.

The existence of similar descriptions of all three forces—electromagnetic, weak and strong—in terms of messenger particle exchanges has encouraged the belief that a common unified description of the forces might be found. Physicists are now confident that the electromagnetic and weak forces are both facets of a common "electroweak" force. Following this success, the merging of the electroweak and strong forces into a "grand unified force" seems a distinct possibility, and although there is no hard-and-fast experimental evidence for this yet, several grand unified theories exist that cast all of these forces into a common mold.

This leaves gravity as the odd man out. To bring gravity into the fold, and produce a fully unified theory of a super-force, it will be necessary to provide a quantum description of gravity. As we explained, the quantum theory began when it was discovered that electromagnetic waves come in dis-crete quanta, or photons; so it seems reasonable to assume that gravitational waves are likewise associated with quanta. These are known as gravitons. As yet, gravitons are entirely hypothetical particles. Indeed, it seems unlikely that their ef-fects will ever be directly observable, so we must rely on the-ory to tell us about their properties. As we mentioned in Chapter 6, gravitational waves travel at the speed of light, so gravitons, like photons, must also travel at the speed of light. But here the resemblance with photons ends. The primary difference concerns the gravitons' weakness in their interac-tion with matter. A beam of gravitons with the same energy and wavelength as a high-power laser (which is a beam of photons) would pass straight through the Earth almost undi-minished, losing less than one percent of its energy along the way. A second difference with photons is that, although gravitons interact with particles of matter only feebly, they will interact with *each other* just as strongly. In contrast, photons, which react strongly with charged particles, do not interact with each other. Two beams of photons will pass through each other unchanged, but gravitons will scatter from other gravitons. Picturesquely, we may say that photons are blind to other photons, while gravitons see and respond to all particles, including other gravitons.

This property of *self-interaction* is at the root of all the difficulty encountered in attempts to formulate a quantum theory of gravity. It is possible, for example, for two gravitons to exchange a third graviton between them, even while the original gravitons are being exchanged between matter particles. With multiple graviton exchanges brought into the picture, it soon becomes horrendously complicated, as we can understand by looking once again at the implications of the Heisenberg uncertainty principle.

Quantum uncertainty allows a messenger particle to come into existence, fleetingly, so long as it soon disappears again. In quantum mechanics, uncertainty is a precise thing, and the energy of the short-lived quantum is determined by the duration of its existence, and vice versa—shorter-lived quanta can have more energy than longer-lived quanta, so that the product of energy and duration is always less than the limit set by the quantum rules.

Because of quantum uncertainty, we can envisage a particle such as an electron as surrounded by a cloud of virtual photons which buzz around it like bees round a hive. Each photon emitted by the electron is rapidly reabsorbed. Photons nearer the electron are allowed to be progressively more energetic because they do not venture far from home, and so need exist only for the briefest duration. Picture, then, the electron immersed in a shimmering bath of evanescent quantum energy, intense near the electron but dwindling steadily with distance. This restless, seething ferment of virtual photons is, in fact, precisely the electron's electric field, described in quantum language. If another electron enters the melee, it can absorb one of the first electron's attendant photons, with the exchange producing a force in the way we have already described. But if no second electron (or other charged particle) is present, the temporary photons have nowhere to go but back to the original electron. In this way, each electron acts on itself through its own photon cloud (Figure 39).

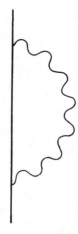

Figure 39. A single electron can emit and reabsorb virtual photons. Such processes produce a contribution to the energy, and hence the mass, of the electron. Embarrassingly, straightforward calculations indicate that this mass "correction" is infinite. This kind of representation (Figures 37–39) is known as a Feynman diagram.

The energy of this photon activity surrounding an electron can be computed. The answer proves, unnervingly, to be infinite. The reason for this apparently absurd result is, however, readily understood. There is no limit to how short a journey a virtual photon may take, and so no limit to how energetic it may be. The contribution from all the nearby photons of unbounded energy to the overall field strength is infinite.

Dodging the infinite
At first sight this bizarre result would seem to imply that the whole theory is nonsensical. But this is not so. Because we can never separate an electron from its attendant photons (we cannot "switch off" its electric charge), there is no way

that this infinite energy can ever be isolated and observed. What we actually observe in the laboratory, and what any other particle in the Universe "sees," is the combined energy of the electron plus its retinue of photons, and this is certainly finite. The infinite self-energy of the electron, while an embarrassing feature of the theory, can, in fact, be sidestepped by deftly dividing both sides of the relevant equation by an infinite amount. Although we were all taught in school to beware of such a step, if it is carried out in a mathematically rigorous manner then consistent results are obtained. To make this still somewhat dubious procedure look respectable, it is dignified with a fine-sounding name—renormalization.

Returning to the subject of quantum gravity, the situation is similar, but worse. Infinities arise in the same sort of way whenever a quantum field process involves a closed loop. Because gravitons can act on each other, graviton loops can be ever more intricate, loops within loops nested like wheels within wheels, and we must suppose that each particle of matter is surrounded by an infinitely complex web of graviton loops. *Every* level of looping adds a new infinity to the calculation, so that as we consider more and more complex processes the infinities accumulate without end.

In QED, the essential trick was to divide both sides of an equation by infinity. The procedure is successful because it has to be done only once. In quantum gravity, by contrast, the equivalent operation has to be performed an infinite number of times. What this means in practice is that almost every calculation carried out using quantum gravity theory in this way produces an infinite answer. The theory has no predictive power, and nobody knows how to extract meaningful quantities from the equations.

The crisis of infinities has been known for decades; recently, though, certain signs have emerged that a way forward may at last be possible. The first clue came, not from the study of gravity, but from the theory of the weak force. For many years the quantum theory of this force was also afflicted by uncontrollable infinities, arising in a similar way, and was useless as a description of all but the simplest types of interactions. In the late 1960s, however, Steven Weinberg and Abdus Salam independently discovered a way of curing the problem. The essence of their approach was an appeal to *symmetry*.

Symmetry has long played an important role in physical theory, and frequently acts as a mathematical guide when the going gets rough. For reasons that are not yet understood (but perhaps may be linked to the cosmic coincidences that make our Universe a suitable home for life), nature conforms to principles that make liberal use of many different forms of symmetry. For example, in the case of most fundamental processes, the laws that govern interactions between particles would be unchanged in a "mirror universe" in which left- and right-handedness were reversed. Again, these laws would be unchanged if past and future were transposed. There are exceptions to these rules (one exception allows the production of the excess of one matter particle for every billion matter-antimatter pairs in the big bang), but by and large the laws of physics are mirror symmetric and time-reversal symmetric.

Most symmetries of interest to physicists are of a more abstract nature; they do not involve geometrical concepts which refer to real space or time. Nevertheless, they still play a crucial role. Abstract symmetries are not always difficult to imagine—for example, there is a type of symmetry between male and female, between positive and negative electric charge and between the north and south poles of a magnet.

These are abstract symmetries that provide a simple relation between physically distinct entities. And by applying abstract symmetries to subatomic particles, physicists have been able to recognize patterns that are not obvious at first sight.

A simple example concerns the proton and the neutron, the stuff of all atomic nuclei. Superficially, these are two distinct particles. The proton is electrically charged; the neutron has no charge and is slightly heavier. Yet in many nuclear processes neutrons and protons behave in identical ways, the charge of the proton acting purely as a label rather than as an extra physical attribute. It is then possible to regard the neutron and proton as merely two *states* of the same basic object, related by an abstract symmetry analogous to that between male and female. Proceeding in this way, many subnuclear particle species have been grouped into family structures, each family representing essentially a single type of object, but possessing several distinct faces.

By exploiting some abstract symmetries in the structure of the weak force, it was possible for Weinberg and Salam to combine the weak force with the electromagnetic force (which possesses a closely similar symmetry structure) and simultaneously to cure the infinity problems of the weak force. This rich bonanza demonstrated that the key to solving such infinity problems in a quantum field theory is to build in as much symmetry as possible and to seek unification with well-behaved quantum fields.

In a head-on attempt to sort out the infinity problems of quantum gravity, theorists embarked in the 1970s on a massive program to exploit the most powerful symmetry yet discovered in nature, known (appropriately enough) as supersymmetry. The essence of supersymmetry is rooted in the concept of spin. Almost all fundamental particles possess a quantum version of rotation, which is called spin and which always comes in certain fixed multiples of a basic quantity. For historical reasons, the basic unit of spin is taken to be one-half. The electron and the neutrino, for example, each have spin one-half. The photon has spin one, the graviton has spin two. No particles are known with spins greater than two, and theory suggests that such objects are impossible.

It is a combination of mass and spin that primarily determines fundamental properties of the different messenger particles, and explains most of the differences between the four forces of nature. The mass of a messenger particle determines the range of the associated force: the bigger the mass, the shorter the range. If the spin of the messenger particle is an even number (or zero), then theory says that the force it produces has to be attractive; if the spin is an odd number, then the force is repulsive.

Nature has made use of massless messenger particles with both spin one and spin two. With no mass, such messengers can range across the entire Universe. The photon is the massless spin-one particle. It does indeed range across the entire Universe, and similar electrical charges (two positive or two negative) do indeed repel one another. The graviton is the massless spin-two particle. It, too, ranges across the Universe, but is always attractive, as theory predicts. There seems to be no force that uses a massless spin-zero messenger, but the theory can tell us what properties such a force would have. It would be a long-range attractive force, similar to gravity, but simpler and without the need to couple universally to all particles of matter.

Gluons behave in a more complicated fashion, and although, like the photon, all eight varieties of them possess spin one, they can interact with each other, and these interactions between different kinds of gluon trap them somewhat and limit their range. The weak force, on the other hand, has a range limited by mass. The W and Z particles are each more than eighty times as massive as the proton, and have a range of less than about 10^{-15} cm.

Although some of this description sounds complicated when expressed in words, in fact nature has a surprisingly limited choice over the properties of the various possible forces, and wherever there is a choice allowed by the mathematics it seems that nature has opted for the simplest alternative, in the sense that it is the one that maximizes the mathematical symmetry.

Before the advent of supersymmetry, particles of different spin value were considered to belong to absolutely distinct families. In particular, all particles with whole-number spin turn out to be force carriers, particles of a quantum field, like photons and gravitons; particles with half-integer spins, like the electron, are always what we think of in everyday language as "real" matter particles. To express this distinction, the former are collectively known as "bosons," and the latter as "fermions." No asymmetry could be more clear, and no systematic connection was known between the properties of bosons and fermions. Supersymmetry changed all that, by providing a mathematical way to link particles of different spin into a single description. It means one can look for laws of physics that transcend the spin barrier and unify differently spinning particles into one superfamily with closely interwoven properties. In particular, it suggests a hidden symmetry between force carriers and matter particles.

Supersymmetry requires that every *type* of particle[1] has a counterpart, with appropriately different spin, in the family of field quanta. Since none of the known "messenger" particles match any of the known matter particles, this demands

---

1 *Not* every individual particle.

the existence of some quantum particles that have not yet been detected, and had not hitherto been suspected to exist. We can make a superficial analogy here with the existence of two families of particles corresponding to matter and anti-matter. The discovery of a mirror-image counterpart to the electron (the positron, or antielectron) required that there should also be an antineutron and an antiproton, to maintain symmetry. In supersymmetry (or SUSY), each known type of matter particle and force field carrier must have an as yet un-known counterpart with a different spin. The discovery of just one of these particles would imply the existence of the whole new family (or families), and as a bonus the theoreti-cal calculations of the properties of these new particles sug-gest that some of them could be just what is required to pro-vide the dark matter in the Universe. So far, however, there has been little direct evidence of a supersymmetric partner to any known particle.

But how is all this supposed to solve the problem of in-finities in quantum gravity? The graviton, which was previ-ously considered to be alone in transmitting the force of gravity, is required by supersymmetry to be accompanied by several types of gravity-carrying messengers called gravitinos, each with spin value three-halves. The existence of gravitinos affects the problem of the infinities. Crudely speaking, the gravitino loops act in a negative fashion, producing negative infinities that, because of the symmetry relations, tend auto-matically to cancel out the positive infinities from graviton loops. As we could never untangle gravitons and gravitinos in the real world, we must always take into account their combined effect. The package of interactions is known as *supergravity*.

Extra dimensions of space

For a while, in the 1970s and early 1980s, supersymmetry seemed to provide the way forward for a consistent theory of gravity within the framework of quantum mechanics; but then it was discovered that the infinity cancellations, although effective for processes involving a few loops, fail when many loops are involved. The setback was short-lived, however, because already an entirely new approach to the problem was being discussed: the possibility that gravity might be unified with the other forces of nature in a mathematically consistent theory once it is recognized that there may exist unseen additional dimensions of space.

The idea that space may have more than three dimensions actually has a long history. Shortly after the general theory of relativity was developed, when only two fundamental forces (gravity and electromagnetism) were properly recognized, a German mathematician called Theodor Kaluza found a way to describe electromagnetism in terms of geometry, just as Einstein had described gravity in terms of geometry. The electromagnetic field, Kaluza pointed out, could be regarded as a kind of space warp, but not a warp in the ordinary three-dimensional space of our perceptions. Instead, Kaluza's space warp lay in a hypothetical *fourth* dimension of space, that, for some reason, we do not see in daily life. If this is correct, one could envisage radio waves and light waves as ripples in the fourth dimension of space. Indeed, if Einstein's theory of gravity is recast in four space dimensions plus one time dimension (making five in all) it yields both ordinary (four-dimensional) gravity *and* Maxwell's equations of electromagnetism. So gravity plus electromagnetism, viewed from four dimensions, is the same as five-dimensional gravity.

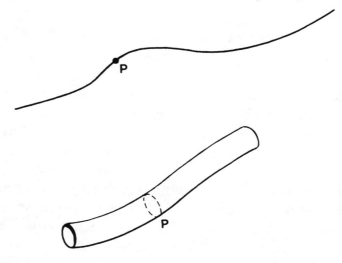

Figure 40. What appears from a distance as a one-dimensional line turns out, on closer inspection, to be a two-dimensional tube. Each "point" on the line is, in reality, a little circle going around the circumference of the tube. In the same way, what we usually think of as a point in space might be equivalent to a tiny circle "going around" a fourth *space dimension.*

Kaluza's theory was taken up by a Swedish physicist, Oskar Klein, who found a way to explain why we do not notice the fourth dimension of space. Klein argued that this is because the extra space dimension is "rolled up." Just as a hosepipe looks like a one-dimensional line from a distance, but is in reality a cylinder, so four-dimensional space could be wrapped into a hypertube (Figure 40). What we previously thought of as structureless points in three-dimensional space are, Klein asserted, really tiny circles in the fourth dimension. The theory even provides a natural circumference for the circle, based on the known value of the fundamental unit of electric charge. The circumference is less than a billion-billionth of the size of an atomic nucleus, so it is no surprise that we cannot directly observe the fourth dimension.

The Kaluza-Klein theory remained little more than a curiosity for several decades. With the discovery of the weak and strong nuclear forces, the attraction of a theory that unified gravity with electromagnetism but left the other forces out in the cold faded. Then, in the early 1980s, the idea that extra space dimensions might exist reemerged. In this modern version of the theory, *all* of the forces of nature are ascribed a geometrical origin. The reason it took so long for physicists to entertain this seemingly logical development of the Kaluza-Klein theory is that while the electromagnetic force requires just one extra dimension for its inclusion within this framework, the weak and strong forces, being more complicated, demand several additional dimensions of space each. To incorporate all the features of all the forces, at least ten dimensions of space, plus one of time, are required.

The proliferation of extra dimensions makes the question of how they can be rolled up to a minute size more complex. It is important that they *are* rolled up, to avoid conflict with observations; but there are many ways in which several dimensions can be "compactified" in this way. For example, two space dimensions can be compactified into either a sphere or a torus. With more dimensions there are more possibilities, increasingly hard to visualize as the number of dimensions increases. In one promising model with a total of eleven dimensions, ordinary four-dimensional spacetime was supplemented by seven dimensions compactified into the seven-dimensional equivalent of a sphere. This is one of the simplest and most symmetrical configurations possible. The seven-dimensional sphere greatly endeared itself to theorists on account of several unique geometrical properties it possesses, some of which were discovered by mathematicians decades ago, long before any physical relevance of such an entity to the real world had been mooted.

It happens that supergravity fitted naturally into this scheme of things. The most economical description of supergravity in mathematical terms turns out to involve precisely eleven dimensions for its formulation. That is, what seems to be a rather intricate set of interrelated symmetries in four dimensions reduces to a single simple and natural symmetry in the mathematics of eleven dimensions. So whether one started from general relativity and the description of forces as a curvature of spacetime, or started from quantum theory and the description of forces in terms of messenger particles, one seemed to be led to the same symmetry in eleven dimensions.

Compelling and elegant though these ideas seemed, the demon of mathematical inconsistency still lurked just beneath the surface of the theory. One difficulty concerned the concept of intrinsic spin. To accommodate spinning particles in the theory, it turns out to be necessary for space and time together to possess an *even* number of dimensions, and eleven, of course, is an odd number. While theorists wrestled with this devastating snag, yet another promising idea came along, incorporating both of the popular concepts of supersymmetry and higher dimensions—and something else as well.

Strings to the rescue?

The essential difficulty with all attempts to provide a unified quantum description of the forces of nature lies with the infinities that always threaten to undermine the predictive power of the theory. These infinities arise, remember, from the fact that messenger particles with ever-higher energy cluster in the regions closer and closer to particles of matter. Infinite quantities occur because there is no limit to how close the messengers can get to the particle of matter that is their source; because the source particles are, in standard theory, mathematical points with zero size, this means that there is no limit to the energy of the closest messenger particles. If, however, the source particle was not actually a point-like entity, with zero size, but was extended in some way, the problem of infinities (which arises, essentially, through dividing an equation by the zero that corresponds to the size of the source particle) would never occur.

Attempts to treat particles such as electrons as little spheres, instead of mathematical points, go back almost a hundred years. These early ideas were not successful because they were inconsistent with the theory of relativity. The novelty of the more recent suggestion is that particles are extended in space in only one dimension. They are not point particles, nor blobs of matter, but infinitely thin *strings.*

Such strings would be the fundamental building blocks of the Universe, replacing the old idea of particles, but resembling particles in that they can move about, while having a much richer dynamical repertoire because of their ability to wiggle as well as to move bodily in space.

In the early 1970s theorists had some limited success modeling the behavior of nuclear matter using string concepts. It seemed in many cases that nuclear particles behaved like whirling string segments. But there were difficulties too. Calculations suggested, for example, that the strings involved could move faster than light, in conflict with relativity theory. For a while, the theory seemed doomed. What saved string theory was the incorporation of supersymmetry. The resulting "superstrings" were properly behaved.

But now another difficulty appeared. The mathematical prescription for such well-behaved strings seemed to include the description of a type of particle that had no place in the known nuclear family: a spin-two particle that had no mass, and therefore moved at the speed of light. This was nothing like the kind of particles involved in nuclear processes; along with its description of the familiar particles and forces, string theory was trying to describe something quite unexpected, which the theorists had not intended to build into the equations. But a spin-two massless particle, though unexpected in this context, is well known to physicists as the *graviton*. String theory was quickly developed as a theory of gravitation. When this was combined with the ideas of supersymmetry, a new type of entity was proposed: the superstring.

It soon became clear that superstrings have some remarkable properties that promise to remove all the troubling infinities that are associated with standard particle theories. At low energies the strings move about as if they were particles, and so mimic all the qualities that have been described so successfully by the standard theories for decades. But as the energy rises to the level at which gravitational forces start to become important, the strings begin to wiggle, and thus drastically modify the high-energy behavior in such a way that the infinities are quenched.

In one formulation of the theory, the strings inhabit a ten-dimensional spacetime; in another version, twenty-six dimensions are required. The ten-dimensional theory incorporates intrinsic spin without difficulty. As in the Kaluza-Klein theory, the extra dimensions are "compactified" to a minute size. Although these extra dimensions can never be "seen" directly, it is intriguing to ponder whether they might be detectable in some way, exerting an influence on our visible four-dimensional spacetime. Quantum physics, as we have seen, links distance with energy. To probe distances one billion billion times smaller than a nucleus requires energy one billion billion times greater than nuclear energies. The only place where such energy was concentrated was in the big bang, where the very early processes that characterized the primeval Universe may, if these ideas are correct, have involved extra dimensional activity in a fundamental way. One attractive possibility is that initially all the space dimensions stood on an equal footing. The denizens of the primeval cosmos—the subnuclear particles—would have "perceived" a multidimensional spacetime. Then a transformation occurred. Three space dimensions swelled up rapidly through inflation to constitute the expanding Universe we now see, while the remaining space dimensions shriveled out of sight, and now manifest themselves not as space but as "internal" properties of particles and forces. Gravity then remains as the only force associated with the geometry of the space and time we actually perceive, but *all* forces and particles are, strictly speaking, geometric in origin.

Strings do not move independently, but can interact with each other, causing them to join together or to split in two. Indeed, the behavior of assemblages of strings is highly complex, and the rules governing their activity are only beginning to be dimly understood. Strings could either be open, with ends waving free, or form closed loops; the most promising approach so far involves loops of string, and seems to contain all of the symmetries that already come out of (or go into) the grand unified theories (known cryptically to mathematicians as E_8), plus those of supergravity as well.

In fact, the full symmetry of this version of the superstring theory actually involves E_8 twice over, in a package that mathematicians refer to as $E_8 \times E_8$. Some theorists have speculated that this duplication involves a sort of second version of the Universe, a shadow world inhabited by identical copies of the sorts of particles familiar in our own Universe (electrons, quarks, neutrinos, and so on) but able to interact with our world only through gravity.

This raises the interesting question of whether we would actually notice the shadow world that interpenetrated our own. It would be possible, for example, to walk right through a person made of shadow matter without feeling a thing. This is because the gravitational force associated with the human body is minute. On the other hand, if a shadow planet were to pass through the Solar System, it could fling the Earth from its orbit. The circumstances would certainly be bizarre, because nobody on Earth would be able to see anything of this celestial interloper; it would be as if some giant unseen hand were scooping the Earth aside.

Looking beyond our Solar System, there might conceivably be shadow galaxies and even shadow black holes. Being purely gravitational entities, the latter would, however, be indistinguishable from black holes formed by the collapse of ordinary matter. If there is a shadow world all around us, though, it would help to explain the existence of some of the dark matter in the Universe. But these are extreme speculations on the fringes of superstring theory. The excitement roused by this theory among physicists has little to do with the possibility of explaining the dark matter, and a great deal to do with explaining how forces combine.

When forces combine

It is too soon to know yet whether superstrings can reproduce known physics and still manage to avoid the infinities that plague more conventional unification theories; but the signs so far are good, even if we may suspect that the more esoteric ideas, such as shadow matter, are likely to fall by the wayside when superstring theory is put on a more secure footing. However that quest is resolved, though, even the established theories of space and time still have room for some further examples of the weirdness of the quantum cosmos, involving the behavior of messenger particles of the cosmic network.

The grand unified theories involve the merging in identity of the different kinds of force. They also involve the merging in identity of different kinds of matter. Ordinary particles fall into two classes: leptons and quarks. The key distinction between them is that only the quarks are subject to the strong force conveyed by the gluons, while the electroweak force affects both. A grand unified force, however, would by its very nature be unable to distinguish between quarks and leptons, since it would include attributes of both the electroweak force and the gluon force.

Calculations suggest that the grand unified force is carried by messenger particles, given the cryptic name X, that possess enormous mass, typically one millionth of a gram—enormous, because this is a million billion (10^{14}) times the mass of a proton. Thanks to quantum uncertainty, virtual X particles can exist for a very brief duration (remember that the lifetime of a virtual particle is inversely proportional to its mass) and so they have only a very short range. Thus an X particle can make a sudden appearance out of nowhere (even *inside* a proton, which has a mass only 10^{-15} that of the X particle it contains!) so long as it promptly disappears again. The appearance can last no more than about 10^{-35} sec. This duration is so short that the ghostlike X particle can travel a mere 10^{-25} cm—about a trillionth of the distance across a proton—before it must give back the mass-energy it has borrowed from the quantum vacuum. With only three quarks present inside each proton, there is very little chance that the X will encounter a quark during its brief life. Very rarely, however, it may happen that two quarks approach each other to within 10^{-25} cm, when there will be just enough time for an X to jump the gap. The chance of such a close encounter has been compared with the random collision of two bees in an aircraft hangar; in fact, to make the analogy work there should be three bees (quarks) in an "aircraft hangar" some 10 million kilometers across. But when such a close encounter does happen, the exchange of the X particle between the quarks has a profound effect, causing the two quarks to change into an antiquark plus a positron.

If this transmutation happens inside a proton, the newly created positron gets ejected, while the antiquark, together with the remaining third quark within the original proton, constitutes a particle known as a pion. After a fraction of a second, the pion itself decays into photons. The net effect is that the proton has disappeared, leaving behind a positron and photons. This means that all matter is unstable: it cannot last forever. The same grand unified theories that provide a mechanism for matter to come into existence also hold the seeds of its demise. Each proton in the Universe is paired with an electron produced in the original matter-creation process. When all the protons have decayed, all the matter which now makes up the stars and planets (and us) will have been turned into electrons and positrons in equal numbers; many of these will in turn meet up, and annihilate into further showers of photons, heralding the ultimate death of matter as we know it.[2] But don't worry unduly about that. This particular version of grand unified theory has not yet been proved correct, and even if it is correct, very close encounters between quarks are so rare that the average proton will take at least 10^{32} years to decay.

How can such a rare process ever be observed experimen-

---

2 Neutrons, except when they are locked up in atomic nuclei, themselves decay after a few minutes freedom into a proton and an electron, so they too suffer the same ultimate fate.

tally? The only way, as we mentioned in Chapter 7 in connection with alpha decay, is to study a very large number of protons for a long time. Because proton decay is a quantum mechanical process, there will always be a tiny probability that any given proton will decay in, say, a year. If the average lifetime of a proton is 10^{32} years, then among a collection of 10^{32} protons there is a good chance of spotting one decay per year. In the early 1980s, an Indian team that had been monitoring a stack of 100 tons of iron with suitable detectors did claim to have spotted proton decay, but the best evidence now is that they were mistaken.

In spite of the lack of direct evidence for proton decay, most physicists believe that the forces of nature do share a common origin at some sufficiently deep level. All the progress in physical theory during the twentieth century has been in the direction of unification—of finding links between hitherto apparently distinct aspects of reality. There is a growing feeling that the physical Universe constitutes a unity binding together not only similar particles in different places, but also the various different particles and forces. Ultimately, one might expect all of nature—particles, fields of force, space and time, and the origin of the Universe—to be part of an all-embracing mathematical scheme. Some optimists, such as Stephen Hawking, believe that this total unification may even be within sight. If that is so, then a mere three centuries of endeavor will have sufficed to transform Newton's cosmic clockwork completely into a cosmic network. But if the task seems easy, some indication of how many mysteries of the Universe still have to be explained can be gleaned from peering into the workings of the ultimate warpings of space and time: black holes.

9 Beyond the Infinite Future

Most people have an instinctive fear of vast spaces. It is an antipathy evidently shared with our ancestors who, appalled at the prospect of an infinite void, preferred to believe that the Universe was confined by concentric shells. Even the idea of space between atoms produces unease. Many Greek philosophers reacted strongly to the assertion by the Atomists that the world consists of particles moving in a void, an antipathy captured by the tag "Nature abhors a vacuum." Even René Descartes declared: "A vacuum is repugnant to reason." Indeed, right up to the beginning of the twentieth century, a philosopher of the stature of Ernst Mach was still rejecting the atomic theory in favor of a material continuum. It seems that the concept of empty space touches on some deep atavistic fears buried in the human psyche. No wonder, then, that people have an awe-filled fascination for the recent speculation that they could be *swallowed up* by empty space.

One of the best-selling science books of all time was John Taylor's *Black Holes,* published in 1973. Although the idea of black holes in space had been shaping up among scientists for some time, they had been given that evocative name only in the late 1960s, and it was only in the 1970s that the general public became aware of them. The bizarre and frightening properties of these objects guaranteed immediate attention, and ensured that the term "black hole" found a permanent place in the English language. These days, it is almost commonplace to read about black holes lurking at the centers of galaxies, busily eating away at the Universe. But only a quarter of a century ago they remained a wild speculation.

Black holes form when gravity, the weakest of nature's four forces, rises to overwhelming proportions. The power of gravity to grow without limit is due to its universally attractive nature and its long range. The other forces are all self-limiting: the nuclear forces are confined to a subatomic range, while electromagnetic forces come in both attractive and repulsive varieties, which tend to cancel out. But go on adding more and more matter to an object and its gravity will continue to rise without limit.

The gravity at the surface of an object depends not only on its total mass, but also on its size. For example, if the Earth were compressed to half its present radius, we would all weigh four times as much. This is because the force of gravity obeys an inverse square law—it gets stronger at shorter distances. The higher gravity would make it harder to escape from a shrunken Earth. At its present size, an object has to be propelled upward from the surface of our planet at about eleven kilometers a second—the so-called escape velocity—before it will fly off into space for good. For a half-size Earth, the escape velocity is about 41 percent greater.

Trapping light

If the Earth could be progressively shrunk (retaining the same mass), the surface gravity and the escape velocity would go on rising. When the compressed Earth was reduced to the size of a pea, the escape velocity would reach the speed of light. This is the critical size. It implies that no light could escape from such an object, so the shrunken Earth would effectively disappear from view. It would become, as far as anyone outside was concerned, completely black. Remarkably, the idea that there might exist objects in the Universe whose gravity is so intense that light cannot flow away from them was mooted some two hundred years ago, by the British astronomer and philosopher John Michell; the Frenchman Pierre Laplace independently hit on the notion of "black stars" a little later.

There is no risk that the Earth will shrink in the manner described: its gravity is safely countered by the solidity of its material. But in the case of larger astronomical objects the situation is different. Stars like the Sun are engaged in a ceaseless battle with gravity. These balls of gas are prevented from collapsing under their own weight by a huge internal pressure. The core of a star has a temperature of many millions of degrees, and this heat produces a pressure sufficient to support the colossal weight of the overlying layers of gas. But this state of affairs cannot last forever. The internal heat is generated by nuclear reactions, and eventually the star runs out of nuclear fuel. Ultimately, the pressure support must fail, and the star will be left to the mercy of gravity.

What happens next depends crucially on the total mass of the star. A star like the Sun will end its days by shrinking to about the size of the Earth. It will then become what astronomers call a white dwarf. Such stars have long been known to exist: for example, the bright star Sirius has a white dwarf companion in orbit around it. Because of the compaction, the surface gravity of a white dwarf is immense. A teaspoonful of the highly compressed matter there would contain about as much matter as a heavy truck does on Earth, but it would weigh 10 million tons in the star's strong gravitational field. White dwarfs avoid further compression with the aid of quantum mechanical effects. Their electrons resist being squeezed any closer together for reasons similar to those that confine electrons in atoms to definite energy levels, and thus prevent normal atoms from collapsing. It is a dramatic manifestation of quantum effects at work.

The ability of quantum effects to stabilize a star was already appreciated by the early 1930s. At that time a young Indian student by the name of Subramanian Chandrasekhar was traveling to England to work with the famous Cambridge astronomer Sir Arthur Eddington. During the long journey by boat he did some calculations and discovered that if a star had about 50 percent more mass than the Sun, then the quantum pressure support provided by the electrons would fail, and the star would collapse further. He showed the calculation to Eddington, who refused to believe it! But Chandrasekhar was right, and heavy stars cannot become white dwarfs.

The further compression of a star, which would occur if it had enough mass for gravity to overcome the resistance of electrons, involves a transformation of the very atomic nuclei that compose most of its mass. The crushed atoms undergo a sort of reversed beta decay, with electrons and protons being squeezed together to form neutrons. At this density the neutrons can deploy the same quantum effects as do the electrons in a white dwarf. So long as the star is not too massive, the upshot is that the star shrinks to become a ball of neutrons packed tightly together, like a monstrous atomic nucleus. As a result of this enormous compaction, the entire star may be no larger than a typical city, yet contain more material than the Sun.

These are the "neutron stars" we have already discussed in Chapter 6. Their surface gravity is so great that the escape velocity is an appreciable fraction of the speed of light. So we know by direct observation, from the fact that neutron stars exist, that there are objects in the Universe close to the "black star" limit discovered by Michell and Laplace.

What about stars so massive that even the neutron support fails? Astronomers are not certain of the exact limit beyond which further collapse must occur, and there is even a conjecture that a still more compact stable phase of matter—a sort of quark soup—might be possible. But a quite general limit on the mass of a collapsed star can be inferred from the theory of relativity. To support a star of a certain mass the material of its core has to have a certain rigidity. The heavier the star is, the stiffer the core material must be. But the stiffness of a substance is related to the speed with which sound moves through that substance: the stiffer the material the faster the speed of sound. If a static collapsed object were as massive as about three Suns, it would have to be so stiff that the speed of sound would exceed the speed of light. As the theory of relativity forbids any physical influence to propagate faster than light, this state of affairs is impossible. The only route left open for the star is *total* gravitational collapse.

If an object implodes from neutron star densities, it will disappear in less than a millisecond, so intense is the pull of its gravity. The surface of the star rapidly crosses the critical radius that prevents light from escaping, so a distant observer would no longer be able to see the object. Although Michell and Laplace were basically right about the possibility of black stars, they incorrectly assumed that the now unseen star could remain static, held up by some ultrastrong force. We now know, from relativity theory, that no force in the Universe can prevent the star from continuing to collapse, once it has reached the light-trapping stage. So the star simply shrinks away, essentially to nothing, leaving behind empty space—a hole where the star once was. But the hole retains the gravitational imprint of the erstwhile star, in the form of intense space and time warps. Thus the region of gravitational collapse appears both black and empty—a black hole.

Starbusting

So much for theory; what about the real world? Astronomers have direct evidence for white dwarfs and neutron stars, but the evidence for black holes is more problematic. What we do have is a plausible scenario whereby black holes might form. The end of a massive star does not come about by wholesale implosion. What happens is a bit more complicated. The nuclear reactions that keep a star hot take place deep in its interior. When the fuel starts to run out, the star can no longer produce enough heat to keep the internal pressure at the level needed to support the weight of the overlying layers. As a result, the core shrinks under gravity. Circumstances can occur whereby this shrinkage takes place in the form of a sudden collapse. When the core implodes in this way, it releases a burst of energy, partly in the form of a shock wave, but also as an intense pulse of neutrinos (neutrinos are among the by-products of the nuclear reactions that go on in the core).

Under normal circumstances neutrinos have almost no effect on matter. They interact so weakly that they pass straight through it. But such is the enormous density of matter reached in the shock wave moving outward from a collapsing stellar core that the flow of neutrinos is seriously impeded. As a result, they exert an appreciable outward push, so that as the core of the star implodes the outer layers are blasted away into space by the released energy. So the star simultaneously implodes and explodes, in an event known to astronomers as a supernova.

Supernova explosions are among the most spectacular events in the Universe. For a few days the shattered star might rival an entire galaxy in brightness, as the explosive energy is converted into light and other forms of radiation. A supernova in our Galaxy is usually readily visible to the naked eye. A famous case is the "guest star" in the constellation of Taurus, recorded by Chinese astronomers in 1054. Today, telescopes show a ragged cloud of expanding gas (known as the Crab Nebula, because of its shape) in the position where the guest star was seen. This is the debris from the death of a star witnessed nearly a millennium ago.

The average galaxy hosts two or three supernovas each century, though none has been witnessed in our own Galaxy since the invention of the telescope. In 1987, however, a supernova was seen in the Large Magellanic Cloud, which is a small satellite galaxy to the Milky Way, visible from the Southern Hemisphere. This event provided scientists with a firsthand opportunity to test their theories about supernova explosions, and the wrecked star has been intensely scrutinized from day one of the explosion. Most significantly, on the day the explosion was spotted visually, a pulse of neutrinos was detected in three experiments (actually looking for proton decay!) around the Earth. Clearly, these were neutrinos from the core of the star, and their arrival on Earth along with the light from the explosion helped to provide dramatic confirmation that our basic ideas about supernovas are correct.

But what about the fate of the imploding core that triggers such an outburst? A study of the Crab Nebula reveals a rapidly blinking pulsar in its midst. Evidently, the core of this particular kamikaze star has ended up as a neutron star. It is easy to believe, however, that a black hole might have formed instead, and most astronomers assume that some fraction of supernova remnants is indeed in the form of black holes.

If a black hole forms as an isolated supernova remnant, it is unlikely that we could detect it from Earth. It is, after all, black. But many stars belong to binary systems, where two stars orbit around one another; if one of these should collapse to a black hole, the other would seem to be orbiting around nothing. In some cases, the tides raised by such a black hole on the surface of the remaining star could strip material away from the star and suck it into the hole. As these gases swirl down into the hole, enormous temperatures are generated, causing intense emission of x-rays. So a good "signature" for a black hole is a binary system which emits x-rays and in which one of the objects is invisible. Several such systems are known. It is even possible to estimate the mass of the unseen companion in one system (known as Cygnus X-1), from the orbital dynamics of the visible star, and to confirm that it is probably above the permissible limit for neutron stars.

The collapse of stars is not the only way in which black holes might form. The more mass that is available, the easier it is to bring about gravitational collapse. For example, black holes with a billion Solar masses of material would form when the density of that material was only about the same as the density of the water in the oceans of our planet. There is some evidence that a black hole with a mass of about a million Suns resides in the center of the Galaxy. Certainly there is a peculiar compact object there which is also a source of intense radio noise and other radiation. Other galactic centers might host still more massive black holes, with the equivalent of a billion Suns of matter. These monsters will betray their presence as material from their vicinity falls into them and is swallowed up. The violence of this engorging releases large quantities of energy that can manifest itself by producing high-speed jets of material or intense bursts of radiation. The galaxy M82 is a good example of such an active system which may harbor a huge black hole.

Another class of objects, the quasistellar sources, or quasars, are associated with disturbed galaxies. Variations in their light show that quasars are only about the size of our Solar System, but they each radiate as much energy as an entire galaxy of 100 billion stars. There is now good evidence that quasars are located at the centers of galaxies, and furnish extreme examples of the kind of activity seen in M82. Many astronomers believe that the central engines powering these objects consist of supermassive black holes immersed in swirling, gaseous material.

By definition, we can never see inside a black hole. But theory can be used to infer what it would be like for an observer to enter a hole and explore its interior. The key to understanding the physics of black holes is the so-called event horizon. Roughly speaking, this is the surface of the hole. Any event occurring within the hole (inside the event horizon) can never be witnessed from the outside because no light (or other form of signal) can escape to convey information about such events to the outside world.

If you should find yourself inside a black hole's event horizon, not only could you never escape but, like the star that preceded you, you would be unable to halt your inward plunge. Just what happens when you arrive at the center of the hole nobody knows for sure. According to the general theory of relativity there is a so-called singularity there, a boundary of space and time at which the original star (and any subsequent infalling matter) is compressed to an infinite density and all the laws of physics break down. It may be that quantum effects cause spacetime to become fuzzy very close to the center, with the singularity smeared out over the Planck scale, about 10^{-35} of a meter. At this stage we do not have a reliable enough theory to know. And it is no good trying to look, or sending an automatic robot probe to look for you. The already fierce gravity of the hole rises without limit as the center is approached, which has two effects. If you fell into the hole feet first, your feet would be closer to the center than your head, so they would be pulled progressively harder than your head, stretching your body lengthwise. In addition, all parts of your body would be pulled toward the center of the hole, so you would be squeezed sideways. At the end of this spaghettification you would be crushed into nonexistence (or lost in a haze of quantum uncertainty). All this would happen in the fraction of a second prior to your reaching the singularity, so you couldn't observe it without being irreversibly incorporated into it.

Where time stands still

Things would look quite different, however, to an observer who watched you fall into the hole from outside. Gravity, remember, not only bends space but also slows time. Near a neutron star the effect is very pronounced, and it is readily detected in pulsar radiation. As you approach the event horizon of a black hole from the outside, so the flow of time in your vicinity slows down more and more as measured by a distant observer. However, the observer who crosses into the hole through the event horizon notices nothing unusual—the event horizon has no local significance—even though at the boundary the time warp becomes infinite. To an outside observer, it will appear to take you forever to reach the event horizon; in a sense, time at the surface of a black hole stands still relative to the time experienced by a distant observer. So anything that happens to you inside the hole lies beyond the infinite future as far as the outside Universe is concerned.

It is for this reason that a journey into a black hole is normally regarded as a one-way trip. To enter a black hole and emerge again would mean that a distant observer would have to see you come out before you went in. In other words, you would have traveled backward in time. This conclusion should not come as a surprise. Because the hole traps light, anything escaping from it has to exceed the speed of light—and, as we have seen, faster-than-light travel can mean backward-in-time travel.

If, then, an object that falls into a black hole cannot re-emerge into the outside Universe, what happens to it? As we have explained, any object that encounters the singularity is annihilated: it ceases to exist. A precisely spherical ball of ordinary matter, for example, collapsing to become a black hole, will shrink to the common center. All the matter will be squeezed into a singularity. But what if the collapsing object is not precisely spherical? All known astronomical objects rotate to a greater or lesser extent, and as an object shrinks, its rotation rate rises. It seems inevitable that a collapsing star will be spinning rapidly, causing it to bulge out at the equator. This distortion will not prevent a singularity forming, but it could be that some of the infalling material of the star will miss it.

Idealized mathematical models of charged and rotating holes have been examined to find out where the singularity lies, and where the infalling material goes. The models indicate that the hole acts as a sort of bridge, or spacetime tunnel, connecting our Universe with another spacetime that is otherwise completely inaccessible to us. This astonishing result opens up the prospect of an intrepid space traveler passing through the black hole unscathed and entering another universe, arriving somewhere beyond our infinite future. If this could be accomplished it would not be possible for the traveler to retrace the passage back to the starting point. Diving into the tunnel from the other universe would bring our daring spacefarer, not back to our Universe, but to a *third* universe, and so on *ad infinitum*. A rotating black hole is connected to an infinite sequence of universes, each representing a complete spacetime of possibly infinite extent, and all connected through the interior of the black hole. The prospect of making any practical sense of this is so daunting that we shall leave it to the science fiction writers.

What would the remote end of the black hole bridge look like to an observer located in the other universe? According to the simplest mathematical models, an observer would see the object as a source of outward-rushing material—the explosive creation of matter—often called a "white hole." Our Universe abounds with exploding objects, such as quasars, and this has led to the conjecture that there are indeed spacetime tunnels channeling matter into our space and time through black holes from other universes. However, few astrophysicists take this scenario seriously. In particular, they point out that the mathematical models neglect the effect of surrounding material and radiation, which would be sucked back into the white hole by gravity, turning it into a black hole. The simple models also neglect the effects of subatomic physics. More elaborate models that incorporate these features indicate that the interior of the hole is so disrupted by these disturbances that the spacetime tunnel would be smashed, and the bridges connecting our Universe with other hypothetical spacetimes would therefore be blocked. The balance of opinion among the experts is that *all* matter entering a black hole eventually encounters a singularity of some sort.

But what if quantum effects remove the singularity somehow? Unfortunately, since we have no complete quantum theory of gravity, we cannot model the consequences of quantum effects in smearing out the singularity with any confidence. Whether they actually result in the complete removal of the singularity is uncertain. Some physicists expect this to be the case, and even argue that the very concepts of space and time will cease to apply under these extreme conditions. Just what sort of structures might replace them is a matter for conjecture. The safest position, therefore, is to regard the singularity as merely a breakdown of *known* physics, rather than the end of all physics.

Wormholes and time travel

The idealized black hole models that permit transit to other universes have been known for over twenty years, and for most of that time it has been assumed that the "tunnels" were mathematical artifacts with no physical significance. Recently, however, the subject has taken a curious twist. A few years ago the American astronomer Carl Sagan wrote a science fiction novel called *Contact,* about an advanced alien community who construct a space tunnel that allows rapid travel between two distant parts of the Universe. In order to make his fictional tunnel plausible, Sagan asked the CalTech astrophysicist Kip Thorne, a black hole specialist, for advice. Intrigued by the idea, Thorne investigated the physics of the proposal with some younger colleagues. The project had a serious side to it too. Thorne wanted to know precisely what restrictions apply to known physics that might prevent such a space tunnel from existing.

Earlier calculations suggesting that black hole tunnels would not be traversable made certain assumptions about the nature of matter. In particular, they assumed that, crudely speaking, matter would always give rise to attractive gravity. But we saw in Chapter 5 how quantum processes can produce *antigravitational* effects under some circumstances. If these circumstances were reproduced in the throat of a black hole tunnel, it might conceivably be possible to evade the "no-go" theorems. The key to antigravity is to produce negative pressure somehow. The CalTech group appealed to the Casimir effect (see Chapter 5) as one example of how a negative pressure and effective antigravity might be produced. They invite us to imagine a pair of reflecting surfaces placed extremely close together. To prevent the surfaces simply sticking to each other as a result of the Casimir attraction, an electric charge is added to each plate in such a way that the electric force of repulsion exactly balances the quantum force of attraction. This peculiar system is then envisaged as lying in the throat of a space tunnel.

The calculations show that Einstein's gravitational field equations can be satisfied by such an arrangement, and that the all-important antigravity of the plate system can be sufficient to counter the tendency of the tunnel to collapse into a singularity. The entrance and exit to the tunnel are no longer strictly black holes, but merely regions of intense gravity which a hypothetical observer could pass through and safely return from without the risk of being swallowed up for good.

The simplest analogy for what might go on is to imagine a journey on the curved surface of the Earth. Suppose you want to travel from London to Adelaide. Because the surface of the Earth is curved, the journey could be made shorter by drilling a hole through the Earth from one city to the other. Then, you could travel in a straight line, arriving in Australia before a colleague who traveled by the conventional route at the same speed that you travel through the tunnel.

It is easy to see how tunnels associated with black holes could do a similar job in a curved spacetime (Figure 41). As usual, we represent spacetime by a two-dimensional sheet, like the surface of a piece of paper. Bending the sheet over makes a flattened U shape, with the top and bottom of the sheet brought close together, and separated by a small gap across the third dimension. If we could join the two opposite parts of the sheet by a tube through the third dimension, it would be possible to travel from one to the other through

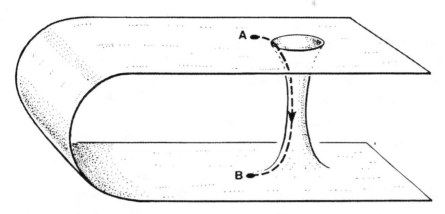

Figure 41. A wormhole joining two otherwise distant regions of space. Traveling through the wormhole provides a shortcut for the journey from A to B.

the tunnel, without going the long way around.[1] Such a connection between different parts of the same spacetime is officially known to relativists as a wormhole. Anything that we can picture happening to a two-dimensional sheet folded through a third dimension can be translated mathematically into four-dimensional spacetime folded through higher dimensions. If the two ends of the wormhole are, say, one light-year apart across the "main sheet," no signal can travel between them in less than a year through that route; but by traveling through the wormhole a signal, or perhaps even a person, can get from one end to the other in much less than a year.

Now imagine the curved spacetime being unbent and laid flat again, with the wormhole stretching and remaining intact. This leaves you with a flat spacetime in which two dif-

1 Since a sheet of paper actually has a finite thickness, you could also simply imagine poking a hole in the paper to make a shortcut from one side to the other.

ferent regions are connected by a U-shaped wormhole, rather like the handle on a teacup. At first sight, this is much less interesting. It looks as if the distance from one end of the wormhole to the other is longer if you travel through the wormhole than if you travel through ordinary space. This is not necessarily the case, however, because space and time behave differently inside the wormhole; even though the spacetime of the parent universe is flat (or nearly so) and the wormhole is curved, it can still act as a shortcut, so that a traveler entering one mouth of the wormhole emerges from the other mouth almost instantaneously, no matter how far away the other mouth is across the universe.

Although the astronautical possibilities opened up by the CalTech proposal are mind-boggling, the really bizarre consequences concern the possibility not of space travel but of time travel. We have mentioned that travel faster than light can be turned into travel backward in time. In going from A to B via the wormhole one is effectively arriving at B ahead of any light signal sent from A along a normal path in space. For example, wormholing from Earth instantly to the center of the Galaxy would get you there 30,000 years before today's sunlight arrived the slow way. This doesn't mean you've traveled 30,000 years into the past, but a slight modification to the wormhole's structure will indeed permit time travel.

The necessary modification is this. One end of the wormhole is kept fixed, and the other end is propelled away at close to the speed of light. If the moving end is then stopped, and returned to lie reasonably close to the stationary end of the tunnel, then a relative time difference will be created between the two mouths of the wormhole. This is a straightforward consequence of the twins effect—the fact that moving clocks run slow, one of the most fundamental and straightforward features of relativity theory, which we discussed in Chapter 3. More time will have passed for clocks next to the stationary mouth of the wormhole than for clocks that traveled with the moving mouth on its journey; so the mouth that has been on the journey will, so to speak, be in the past of the mouth that stayed put. But the "present," for anyone traveling through such a wormhole, always corresponds to the time at the mouth by which the hole is entered. It follows that if you were to enter the wormhole from the end that has been moved you would emerge from the fixed mouth of the wormhole to find that you had traveled backward in time. Provided that the two ends of the hole are close enough together as measured across normal space, you would be able to complete a whole circuit and return to your starting point *before* you jumped into the hole. By passing repeatedly through the wormhole, you can move farther and farther back in time—but you cannot travel back to times before the moment when the moving mouth began its journey and the time dilation effect began to work.

Hardly surprisingly, there are many caveats to this scenario. One concerns the nature of the all-important reflecting surfaces that produce the Casimir effect. It is essential that their own mass and internal structure do not add more to the gravitational force than the antigravity that they produce, and it is hard to see how this could be achieved in practice. Moreover, some way has to be found to allow the putative time traveler to pass through the reflecting surfaces (a trapdoor?) without upsetting the delicate equilibrium of the system. Another problem concerns the manipulation of the wormhole ends. Being simply empty (albeit curved) space, the ends of a wormhole cannot be seized physically, with a pair of giant tongs, and accelerated like a lump of matter. Some sort of electrical or gravitational force would have to be exerted on the wormhole ends—and the whole operation must be carried out without the diameter of the wormhole shrinking to zero while it is being stretched and returned to its former condition. Quite apart from all these problems, there is the question of creating the wormhole in the first place.

Now, we want to emphasize that none of these wormhole speculations are intended as serious practical suggestions. They fall into the category of thought experiments, like Einstein's and Schrödinger's speculations about the nature of quantum reality, designed to test the consistency of the laws of physics. The conventional position is that time travel should not be permitted by any physical process whatsoever, precisely because it would threaten the consistency of physics.

Consider, for example, the case of the time traveler who visits his grandmother when she is still a child and murders her. If the grandmother dies as a child, then the time traveler could never have been born, and could not, therefore, carry out the murder after all. But if the grandmother was *not* murdered, the time traveler *would* have been free to murder her. . . . This type of internal contradiction can be removed only by invoking some strange additional features for physical theory. For example, there may be a law of physics that says only self-consistent causal loops can exist, so if anyone attempts the granny-murdering "experiment," then something would always happen to frustrate it—the would-be murderer's gun will misfire, or it will turn out that he was adopted, or whatever it might be. Or if one believes in the many-universes theory, perhaps all such time-travel scenarios involve altering the past, not of the world from which the time traveler comes, but of a closely similar parallel world.

Whatever oddities may be implied in these "back-to-the-future" experiments, it is important to investigate whether the known laws of physics alone rule out time travel, or whether some additional principles are involved. This is the real motivation for the work of Thorne and his colleagues.

In fact, the topic of wormholes is currently being intensively investigated by many different research groups, but *not* to test the time-travel hypothesis. Interest focuses instead on the properties of the microscopic virtual wormholes that we mentioned briefly in Chapter 5—the ones that occur naturally amid the spacetime foam. Just as quantum fluctuations in the vacuum create temporary photons, so, on an even smaller scale, they should spontaneously create temporary (virtual) wormholes. The size of these wormholes is typically twenty powers of ten smaller (10^{-20}) than an atomic nucleus. Thus, on an ultramicroscopic scale, space would be a labyrinth of such structures, endowing it with the complicated topology that has been dubbed spacetime foam. With masterly understatement, such tiny tunnels through spacetime are simply referred to by relativists as "microscopic" wormholes.

Would-be time travelers speculate that if a virtual microscopic wormhole were pulled out of the spacetime foam and stretched to macroscopic proportions, then it could be used as a time machine—they suggest that the very space around us is populated with a vast number of tiny, short-lived, natural time machines. "All" we need do is to learn how to harvest them. But capturing and expanding a wormhole, not to mention shoring it up with exotic matter against collapse, are, we repeat, not a serious practical proposition. What is taken seriously is the possibility that virtual quantum wormholes might offer a clue to one of the great outstanding mysteries of modern physics.

What is the weight of empty space?

The idea that empty space can have weight might appear fanciful, or even meaningless. How can "nothing" weigh something? It must be remembered, however, that empty space is far from being nothing. Even when all particles are removed from a region of space, it will still contain virtual particles, temporarily created from quantum processes, and bestowing upon the vacuum both energy and pressure. This energy is associated, through Einstein's $E = mc^2$ relation, with mass; and one might expect this mass to gravitate.

Unfortunately, finding out does not simply involve putting an empty box on a set of scales and weighing it. Space is all around us, and if it does gravitate it will pull equally in all directions. The only way this gravitational pull will manifest itself is in the motion of the Universe as a whole. In Chapter 5 we discussed how the effect of the quantum vacuum energy is to cause not gravity but *antigravity,* because the associated pressure of the vacuum activity is negative. According to the inflationary scenario, this "negative weight" of space caused the brief but violent period of accelerating cosmic expansion during the very early stages of the Universe.

Following the end of inflation, the weight of space became essentially zero. Attempts have been made, however, to detect any tiny residual effect that may remain in the Universe today. If the weight of space remained, by even the most minute amount, different from zero, this would show up in the way the Universe expands, by competing with the way that the attractive force of ordinary gravitating matter gradually slows the expansion.

No such effect has been detected, and a limit can be placed on the maximum possible weight of space. It is a staggering 120 powers of ten less than the value that, theorists calculate, prevailed during the inflationary phase. The number 10^{-120} is so incredibly small that it is tempting to believe that the weight is now truly zero. But this conclusion presents something of a mystery. We would expect the energy of the quantum vacuum to be typically very large—of inflationary proportions, in fact. So we have the curious situation that the inflationary phase seems to be the "natural" state of affairs, while the close-to-zero present-day value of the weight of space seems to be peculiar—even contrived.

Why contrived? The accuracy of the term becomes clear when we try to understand *how* the present value could be so small. Quantum vacuum energy can, in fact, be either positive or negative depending on the nature of the fields involved. If nature could arrange for the negative and positive contributions to cancel out, the result would be zero. But that would require a Cosmic Accountant with an exquisite skill at balancing the books. In order to get rid of this unwelcome contribution, the influences of different kinds of particle and fields have to be delicately arranged to cancel against each other to at least a precision of 120 decimal places. It seems highly implausible that this would occur accidentally. The only alternative is for there to be a natural mechanism that *forces* the weight of space to be zero.

This is where wormholes come in. One of the fields that contribute to the quantum vacuum energy is the gravitational field, and it is quantum fluctuations of the gravitational field that create not only baby wormholes but other distortions in the geometry of spacetime. Some of these distortions will take on the form of entire "baby universes" connected to our own spacetime by a wormhole, as if by an umbilical cord. The whole process takes place on an ultramicroscopic scale, and one must imagine these tiny protuberances continuously fluctuating, sometimes disconnecting themselves entirely from our Universe as their umbilical wormholes pinch off, sometimes being reabsorbed back into our spacetime as the quantum fluctuation fades away. The cumulative effect is to clothe the space of our Universe with a sort of gas of shifting minispace bubbles. But each minispace is its *own* space and time, just like the parallel universes we discussed before. The only connection with our spacetime is through the umbilical wormholes, and the diameters of the wormhole mouths are, remember, far smaller than the diameter of an atomic nucleus, so we cannot observe them directly.

How will all this affect the nature of the vacuum? The task of computing the effect of this monstrously labyrinthine spacetime froth on the weight of the space to which it clings has been taken up by Stephen Hawking of Cambridge and Sidney Coleman of Harvard. Their calculations appeal to a universal principle of physics, known as the principle of least action. It states, roughly, that whenever anything changes, it does so in such a way as to minimize the effort. For example, a pool ball always chooses to travel along a straight path between two points rather than exert itself by following a zigzag path, unless it is acted upon by external forces. This law of natural indolence, when applied to wormhole fluctuations, implies that those baby universes with very little vacuum energy are preferred to those with a lot. The most preferred are those with precisely zero energy, so the appropriate quantum "average," or expected, value of the vacuum energy will be very close to zero, and this value permeates our Universe from the myriad baby universes to which we are connected.

If these calculations hold up, we will have arrived at a curious conclusion. Our naive expectation that empty space is weightless turns out to be correct, but not for the reasons we thought. It has nothing to do with its emptiness as such, because even empty space is alive with quantum activity. The weightlessness is due instead to an unseen froth of parasite universes that cling to our spacetime through a network of invisible wormholes. Without wormholes the Universe would collapse.

The "weighty" issues raised in this section demonstrate vividly how the old Newtonian mechanistic paradigm has been superseded. Far from dominating the activities of the cosmos, matter seems to assume an almost peripheral role. The main activity comes instead from the most insubstantial entities conceivable, a foam of fleeting quantum wormholes, nothing more than a froth of empty space whipped into half-real tunnels, knots and bridges. And it is only by leave of the special properties of this foam that ordinary matter exerts the influence it does in the Universe today; for had the weight of space not been so incredibly close to zero, it would have been quantum vacuum energy, not the gravitation of matter, that determined cosmic dynamics.

In the past few chapters, we have shown how the quantum and relativity revolutions have transformed our image of nature from that of a clockwork to something far more subtle and nebulous. But these transformations pale in significance beside the impact of the new information revolution. As we discussed in Chapter 2, scientists are increasingly thinking of the physical Universe less as a collection of cogs in a machine and more as an information-processing system. Gone are the clodlike lumps of matter, to be replaced instead by "bits" of information. This is the shape of the emerging universe paradigm—a complex system in which mind, intelligence and information are more important than the hardware. The time has come for us to take a look at life, mind and intelligence, not as a parochial human concern, but in their cosmic context.

10 The Living Universe

Many ancient cultures believed that the Universe was a living organism. Aristotle, who had a deep interest in biology, was impressed by the way that living things seem to be motivated by purpose, their actions forming part of a plan directed toward some predetermined goal. When we see a bird building a nest, for example, it is clear that its behavior is related to the concept of laying eggs and caring for its young. Whether the bird has any conscious awareness of what it is doing is more contentious, but its activities are certainly not random, and can be properly explained only by taking into account the end product.

It is tempting to generalize from the biological realm and to bestow a purpose on all of nature. People often use purposeful language informally, as when describing water "seeking its own level" or the weather "trying to improve." The idea of matter being an active agent, rather than an entity that is passively pushed and pulled by blind forces, appeals to something deep in our makeup. Notice how easily children accept stories of inanimate objects—such as buses, trains and even rocks or clouds—as living things, with personalities and emotions. According to Aristotle, the entire Universe resembles a gigantic organism, and is directed toward some final cosmic goal. The idea that physical processes can be determined by, or drawn toward, a predetermined end state is known as teleology.

With the rise of modern science, and in particular the Newtonian mechanistic paradigm, teleology was abandoned (at least outside of biology) and replaced by the concept of the cosmic clockwork. And yet, in these supposedly mechanistic and rational times, one of the few ideas to emerge from science and reach out to strike a chord with a wide cross-section of ordinary people in recent decades has been the concept of Gaia, the hypothesis that the Earth itself may, in some sense, be regarded as a single living organism.

No issue confronts the paradigm conflict we have been describing so starkly as the mystery of life. According to the mechanistic point of view, living organisms are just machines, albeit complicated and wonderful machines. The evolution of life on Earth is likewise seen as a mechanistic affair, but a creative element is introduced through random variations. Most biologists accept that random mutations and natural selection alone can satisfactorily account for the form of all living organisms, once life had got started. As regards the origin of life, this is more problematical. It is usually supposed that the precise physical processes leading to the first living organism were exceedingly improbable, and in any case they remain shrouded in mystery. From this point of view, life may well seem to be unique to planet Earth, since the sequence of events that led to the first organism would be unlikely to have been repeated elsewhere.

In contrast to this philosophy, the new viewpoint recognizes the creative and progressive nature of most physical processes. No sharp division is drawn between living and nonliving systems. The origin of life is regarded as just one step (albeit a significant one) along the path of the progressive complexification and organization of matter. If matter and energy possess an innate tendency to self-organize, then one would expect to find life arising again and again, given the right conditions. In that case we might expect there to exist other planets with life, and possibly even intelligent beings. The discovery of life elsewhere in the Universe would thus lend powerful support to the postmechanistic para-digm—provided, of course, that it could be shown that the "alien" life had indeed arisen independently.

Recently, advances in space technology have permitted the first rudimentary systematic search for extraterrestrial life. The issues involved are fundamental in shaping our perspective of ourselves and our place in the natural world, as well as having a direct bearing on the need for a new paradigm. But before we can begin to search for extraterrestrial life, we must have a clear idea of exactly what we are looking for. What, in fact, is life?

What is life?

We have no difficulty in recognizing life when we encounter it on Earth. Men, mice, mushrooms and microbes are all un-deniably living. Yet what essential features do these systems display in common? Frequently cited properties of life are the ability to reproduce, response to external stimuli, and growth. The problem here is that other, manifestly inanimate systems also share these properties.[1] Flames readily repro-duce. Crystals both reproduce and grow into more organized structures. Bubbles respond to external stimuli by retreating when approached.

Furthermore, once we probe below the level of our every-day experience—below the level accessible to our senses, especially those of sight and touch—there is no clear divi-sion, after all, between what is living and what is not. The classic example is the virus. In spite of the fact that viral dis-eases clearly involve biological activity, the virus itself does not even satisfy one of the criteria for life we have already

1 Which, in fact, is another expression of the principle of self-organization of complex systems, both living and nonliving.

mentioned—it cannot reproduce by itself, or with the aid of another virus. A virus can multiply only by invading a host cell and taking over its biochemical functions. In essence, it turns the cell into a production line for more viruses. It could be argued that under these circumstances the *cell* is no longer living, since it has lost the ability to reproduce *itself*. In isolation, though, viruses can be reducéd to an inert dry powder, and differ little in their properties from other substances with less organized biological effects.

These difficulties oblige us to adopt a rather vague definition of life. Certainly a high degree of organization is a necessary requirement. Probably we should avoid thinking in terms of individual organisms at all, and direct our attention instead to the complex interdependence of vast numbers of different life forms. On Earth, this is called the biosphere. It is doubtful if any particular organism could survive in isolation on Earth; only the total intricate network is viable.

This brings us, by a different route, to the controversial concept of the various forms of life on Earth as components of one living organism, which is the basis of the Gaia hypothesis. Jim Lovelock has pioneered the idea, which has stirred fierce debate among biologists and ecologists, but has become very fashionable in some quarters, often with embellishments that Lovelock himself disowns. We do not have space to digress into a detailed discussion of that debate here, but we do wish to point out that the concept of Gaia fits naturally into the new paradigm of self-organizational complexity. Not only that. If the living forms on Earth are seen as components of a single more complex system, whether it be called "the biosphere" or "Gaia," it is reasonable to conjecture that during the long future evolution of the Universe the growth of complexity may develop to embrace not just individual planets but entire star systems and ultimately, if there is time enough, whole galaxies, in a living cosmic web of interdependence. But all that lies in the very far future; we are more concerned at present with the past, the other end of the chain. How did life on Earth begin?

Since Darwin, biological thinking has been dominated by the concept of gradual evolution. From the fossil record one may infer that the present condition of the Earth's biosphere is the product of an immense number of successive steps toward ever greater complexity, adaption and sophistication. For example, 500 million years ago there were no living things on the land. About 2,000 million years ago there were no creatures with backbones. The oldest rocks of all, dating from around 3.8 billion years in the past, contain traces of only the most elementary microscopic life forms. Given this progression from simplicity to complexity, together with the existence of viruses that seem to bridge the gap between the living and the nonliving, it is tempting to conjecture that the origin of life on Earth was simply another step in a general evolutionary sequence, part of the pattern of cosmic self-organization. So can living matter be created unaided out of inanimate chemicals?

The origin of life

The idea of the spontaneous generation of life has a long history. A favorite recipe, not many human lifetimes ago, was a piece of rotting meat from which maggots would eventually be seen to appear "spontaneously." But this is not what we now mean by the creation of life out of nonlife. The work of Louis Pasteur finally demolished these naive ideas, and today the study of the spontaneous generation of life belongs firmly in the realms of biochemistry.

A realistic attempt to investigate the generation of life on Earth was carried out by Stanley Miller and Harold Urey at the University of Chicago in 1953, in an experiment that is now regarded as a classic of its kind. Miller and Urey hit upon the idea that if the conditions believed to have prevailed on the primeval Earth were reproduced in the laboratory, the first steps toward the chemical synthesis of living material might be induced to recur. In accordance with current ideas of the time, they filled a flask with methane, hydrogen, ammonia and water, thought to represent the composition of the Earth's atmosphere in the remote past. The present atmosphere of (chiefly) oxygen and nitrogen appeared on Earth later in its evolution, and is itself a result of biological activity—a sign, to any extraterrestrial community with powerful enough instruments to detect the presence of these gases in the atmosphere from a distance, that the Earth is a home for life.

The Miller-Urey experiment, which lasted several days, involved passing an electric spark through the chemical mixture, simulating the energy input from thunderstorms in the atmosphere of the primitive Earth. The liquid slowly turned red, and when it was analyzed it was found to contain substantial quantities of organic molecules[2] known as amino acids. Amino acids are *not* living molecules themselves, but they are the building blocks of proteins, which are essential components of living things on Earth. Inside the cells of your body, coded messages in the DNA are translated by RNA into working molecules of protein, which carry out the functions

2 Organic molecules are molecules that contain carbon, an atom with a unique ability to form many highly complex molecules in association with other atoms, especially including those of hydrogen. Such complex molecules are associated with living things, hence the name; but they can also be produced in other ways, so that although organic molecules are essential for life as we know it, the existence of organic molecules is not on its own proof of the presence of life.

of life. To some, it seemed that Miller and Urey were already, in the early 1950s, on the road to creating life in the laboratory. Admittedly, there is a big step from a collection of amino acids to the first replicating organism, yet given the millions of years available since the formation of the Earth, one can perhaps envisage the "soup" of amino acids gradually becoming more and more complex, as the organic molecules continually jostle with one another and stick together in various ways.

Unfortunately, it is not that simple—for reasons that, getting slightly ahead of the historical development of the story, we have already hinted at with our mention of DNA. It was not until 1953, the same year that Miller and Urey first carried out their famous experiment, that Francis Crick and James Watson, working in Cambridge, established the structure of DNA, the famous double helix, paving the way for further studies which established the mechanism by which all Earth life works. Until then, there was a respectable school of thought which held that proteins were the key to life, and therefore that by creating amino acids the secret of life might be unlocked; after the importance of DNA became clear, the significance of the discovery of a natural amino acid factory had to be demoted.

All Earth life is ultimately dependent on these two groups of chemicals, nucleic acids and proteins. Both are made mainly from carbon, hydrogen and oxygen, together with small quantities of other elements such as sulfur and phosphorus. Proteins are built out of about twenty different types of amino acids in different combinations (not all twenty amino acids in every individual protein). Proteins have a dual role, as structural elements and as catalysts (known as enzymes) that greatly enhance the rate of crucial chemical processes. Without enzymes, life would simply grind to a halt.

Nucleic acids are responsible for storing and transmitting all
the information required to build the organism and make it
function—the genetic code. The code includes instructions
for the manufacture of specific enzymes and specific structural proteins. One type of nucleic acid, DNA (short for
deoxyribonucleic acid), takes the form of the now familiar
long-chain molecules wound into a double helix. The double
helix is where the information needed to replicate and operate the organism is encoded.

In familiar inorganic substances, such as air or water, a
typical molecule will consist of two or three atoms bound together by electrical forces. In contrast, a molecule of DNA
may contain many millions of atoms. Indeed, each cell in
your body contains enough DNA to stretch, laid end to end,
over a distance of 180 centimeters. The arrangement of
all the atoms along these chains is not just random, but
intricately organized in a highly specific way. Changes in
the ordering of certain crucial DNA subunits will make
the difference between an elephant and a flea or,
more subtly, between you and a chimpanzee. The bewildering variety of life forms on Earth is an indication of the
enormous number of available combinations of these subunits.

In fact, the number of possible ways in which atoms of
carbon, hydrogen and oxygen could form into molecular
chains the size of the DNA molecules in your cells is incon-
ceivably large. The probability that a molecule as complex
and specific as the DNA that codes for a human being would
form purely at random from a soup of simple organic subunits is negligibly small. If that was what had actually happened, then life would indeed be a miracle.

But what about Darwinian variation and natural selection?
Cannot that process alone be responsible for generating the

complexity of DNA and proteins? Unfortunately, traditional evolutionary effects are of little help in steering the prebiotic "soup" toward truly living material. The concept of the fittest, best-adapted organism possessing a selective advantage over its rivals, and therefore surviving to populate the environment with more of its progeny than they do with theirs, hardly applies to an inanimate molecule incapable of replication anyway.

Little is known about the crucial jump from amino acids to proteins, and even less about the origins of nucleic acids. It is conceivable that some variant of the Miller-Urey primeval soup would, if left long enough, find itself gradually directed toward the "right" sort of molecular arrangements automatically.[3] For example, the action of randomly formed enzymes would lead to the high concentration of certain types of molecules at the expense of others. If those molecules in turn tended to form the very enzymes that help produce them, then a self-reinforcing cycle would arise. Whole successions of interlocking cycles could then raise the level of complexity stage by stage until, eventually, the first giant molecule capable of reproduction would be synthesized. Thereafter the going gets easier, as this fertile molecule sets about converting the remaining contents of the soup into replicas of itself. The way is then thrown open for Darwinian evolution to get to work.

Was this how life began on Earth? That is what many scientists claim. If they are correct, then it seems that the spontaneous generation of life from simple inanimate chemicals

---

3 One speculation is that early steps toward life occurred when molecules were not roaming free in a liquid, where they might only briefly chance to meet (and where collisions would tend to break apart complex structures), but were held in place on the surface of a claylike solid, acting as a template and allowing time for interactions between neighbors to build up complex chains.

occurs far more easily than its stunning complexity would suggest. Earth is only about 4½ billion years old, and for several hundred million years after the planet formed, bombardment by huge meteorites and high temperatures would have destroyed any faltering steps toward life. Yet fossil records date life on Earth from at least 3.8 billion years ago. It seems that no sooner did the Earth become habitable, than primitive life appeared. To some scientists this promptness has suggested that life is an automatic and inevitable consequence of appropriate physical conditions—an alternative phase of matter that arises naturally given the right raw materials. If they are right, then it is clear that, far from being miraculous, life is a rather common feature throughout the Universe. So—where is it?

Worlds beyond

Since the time of Copernicus, almost five hundred years ago, humankind has had to learn and relearn the salutary lesson that there is nothing special or privileged about the Earth. It is a typical planet near a typical star in a typical region of a typical galaxy. Can we suppose that Earth life is an exception to this "principle of terrestrial mediocrity"? Or should we, in the spirit of Copernicus, argue that life is also a typical product of a planet like the Earth?

If life does automatically form under the right conditions, our quest for extraterrestrial life turns on the search for other worlds in the Universe on which these conditions are likely to be fulfilled. Given another Earth-like planet elsewhere in the Galaxy, some form of life would, according to this point of view, eventually arise there. But a search in our immediate neighborhood of space is not encouraging. Our eight sister planets in the Solar System all differ from Earth in conspicuous and possibly lethal ways. Nevertheless, they are not a complete write-off.

Mars has long been regarded as the most likely candidate for Earth-type life in our Sun's family of planets. The Martian climate is hardly equable by Earth standards—it is intensely cold and the planet has an extremely thin atmosphere—yet some forms of Earth life survive on our planet under equally harsh conditions, and would, indeed, probably be able to live on Mars if they were transported there. Moreover, there is evidence that substantial bodies of liquid water—which is an essential ingredient of life on Earth—existed on Mars sometime in the distant past.

It is important to remember that life has evolved on Earth into a wide variety of forms, each beautifully adapted to the physical conditions of its own ecological niche, although those physical conditions may differ markedly from one part of the globe to another. For example, bacteria are known that survive in the boiling eruptions from geysers, while other microorganisms survive in the intense cold of Antarctica, where conditions are not so different from those on Mars. Even if it were not true that conditions on Mars today were able to support some forms of existing Earth life, it would still be possible that life had developed on Mars during the earlier wet phase of the planet's evolution, and had subsequently adapted to flourish in the modern Martian conditions—conditions that human beings regard as hostile.

Mars was, in fact, the subject of a detailed life-detection investigation, part of the two Viking lander space probe missions, in the 1970s. Four separate experiments attempted to detect the effects of organisms (like those that live on Earth) in the Martian soil. One experiment yielded positive results, another was negative and two gave unexpected and puzzling results. A single negative result to such an experiment does not on its own mean that there is no life on Mars, only that the experiment failed to detect it.[4] A single positive result *ought* to mean that there is life on Mars, but in view of the ambiguity of the other experiments and the possibility that (since none of the other experiments found life) there might have been some shortcomings in the design of this particular experiment, the results should not be taken at face value, and most scientists are cautious. They will go so far as to say only that there seems to be some *unusual* chemistry at work in the Martian soil, without saying that this must be *bio*chemistry. So, in the light of the results of the Viking experiments, the issue of life on Mars is still open, although it is clear from the pictures sent back by Viking that, at least in the vicinity of the spacecraft, there are no large plants or animals.

Perhaps the best hope for life elsewhere in our Solar System now rests with Jupiter and with the huge moon of Saturn known as Titan, both the subjects of investigation in the 1980s by the Voyager spacecraft. Many researchers believe that conditions on Jupiter, although very cold, resemble chemically those of the primeval Earth. In a sense, the atmosphere of Jupiter—rich with gases such as methane and ammonia, with huge storms swirling through it—is a sort of gigantic Miller-Urey experiment. Its multilayered structure provides a whole range of different chemical and physical conditions, some of which ought to suit primitive life, and even the coloring of some of the bands of Jupiter, red and reddish-yellow, is the same as the color of the products of the Miller-Urey experiment.

Titan, though found to be disappointingly cool, has a dense atmosphere of nitrogen and could even possess liquid nitrogen seas. It resembles a supercooled version of the

4 An elephant trap set up in downtown Quebec might fail to trap any elephants, but that would not be proof that there are no elephants on Earth.

prebiotic soup, put in cold storage when the Solar System formed more than 4 billion years ago. In another 4 billion years or so, however, the Sun will, according to well-established astronomical theory, grow larger, swelling up to become a red giant and radiating more heat. Will Titan then be brought out of the deep-freeze and warmed into a state resembling that which proved ideal for life to emerge on Earth? Perhaps we are distanced from life elsewhere in the Solar System more by time than by space.

The other planets of the Solar System are even less promising abodes for life than Mars, Jupiter or Titan. The real hope for Earth-type life must rest, for now, with the stars. Our Galaxy alone contains 100 billion other suns, many of which could be accompanied by planets similar enough to Earth to make them suitable abodes for life. As even the best telescopes on Earth cannot directly detect these other Earth-like planets (although there is hope that the orbiting Hubble telescope may be able to do so, once its initial problems have been rectified), this assumption rests on theoretical arguments only. Though opinions differ as to the quantity of Earth-like planets, and precisely how close to terrestrial conditions a habitable planet needs to be, the numbers involved are so large that it would be surprising indeed if there were no other suitable planets in the Milky Way Galaxy; even if only a fraction of one percent of all stars were accompanied by a family of planets like our Solar System, there could be hundreds of millions of planets in the Milky Way suitable for life as we know it. And there are millions of other known galaxies. . . .

This kind of speculation is, however, open to the charge of extreme chauvinism, and may err on the pessimistic side. Why should alien biology conform to the tightly restrictive principles that govern terrestrial life? Perhaps life can exist in countless other ways that do not involve proteins or nucleic acids at all.

DNA is but one of an almost limitless variety of alternative long-chain molecules based on carbon chemistry. Who can guess what other arrangements may be possible? Can we really suppose that such a specifically intricate structure as DNA would be the only route to biology? And what about alternative chemical elements? Silicon, for example, although not quite so versatile as carbon, can perform a similar chemical function. Such is the immense variety of energy sources and chemical reactions available that untold possibilities suggest themselves. But precisely because it is all speculation, exotic biologies cannot be taken too seriously. The great merit of carbon chemistry and biology based on DNA as the archetype in our search for life elsewhere in the Universe is that we *know* the system works here on Earth.

If, though, life does exist elsewhere in the Universe based on alternative chemical processes, then it could flourish in the most bizarre environments. Colorful images have been constructed of organisms wallowing in the liquid nitrogen seas of Titan, or crawling across the baking deserts of Mercury. Beyond the Solar System, billions of other planets of wildly different forms could play host to all sorts of weird and wonderful creatures. In fact, accepting alternative chemistries, it is hard to envisage planets where some form of life could *not* flourish. After all, the underlying physical principle that seems to be involved in self-organization and complexity, up to and including biological complexity, is simply the requirement of an open system through which energy and entropy flow, and a suitable energy source (which often simply means a temperature difference) to "feed" off.

Life without worlds

A few scientists have even gone beyond the notion of alien chemistry and proposed the idea that life elsewhere could be based, not on chemistry at all, but on some other complex physical process. A famous example, developed by astrophysicist Fred Hoyle in his science fiction novel *The Black Cloud,* envisages a huge cloud of tenuous interstellar gas organized into a thinking, purposeful individual, moving from star to star to feed off the free energy available.

In recent years Hoyle has built a detailed theory whose roots can be traced back to this idea. In collaboration with Chandra Wickramasinghe, he now suggests that the microscopic grains of interstellar material found within such interstellar clouds (and studied by astronomers using infrared telescopes) are in fact living bacteria encased in a protective shell. Hoyle and Wickramasinghe challenge the traditional assumption that life as we know it began on Earth, and revive an old theory developed almost a hundred years ago by Svante Arrhenius, a Swedish polymath who, among other things, carried out some of the first detailed calculations of the greenhouse effect. Arrhenius proposed that life might be spread across the Galaxy in the form of microorganisms riding on dust particles pushed around by the pressure of starlight. In the Hoyle-Wickramasinghe version, an enormous number of different microorganisms pervade interstellar space, ready to be swept up by any suitable host body, such as a planet or a comet. This could neatly explain how life became established on Earth so quickly after the formation of the planet, and the implication is that any similar planet would be similarly infected with life equally promptly. By providing billions of years for prebiotic chemistry to have been at work on the material in clouds between the stars before the Earth ever formed, the theory makes the whole business of life arising out of nonlife by chance that much more credible.[5] It is hard to place much credence, however, on the still more speculative notion put forward by Hoyle and Wickramasinghe that our planet is being continually "reinfected" by microorganisms from space, which are responsible for major epidemics of diseases such as influenza. A key test of the Arrhenius-Hoyle-Wickramasinghe idea is the existence (or nonexistence) of life on Mars. Since that planet is a prime candidate for infection in this way—and it is hard to imagine microorganisms that could survive the rigors of interstellar space being unable to get a foothold there—the continuing failure positively to identify life on Mars must count against this theory.

How, though, can extraterrestrial life be discovered, if the rest of the Solar System is barren? Our space probes are unlikely to reach other stars in the foreseeable future. Should our sister planets in the Solar System turn out to be sterile, will the subject of life beyond the Earth remain solely within the realms of science fiction? Perhaps not, since there may be another way to test the conjecture that we are not alone.

Aliens at large

Although the discovery of the smallest extraterrestrial microbe would alter for ever humankind's perspective on the Universe, the real fascination surrounds the possibility of other *intelligent* life-forms and alien technological communi-

5 More credible *in principle,* because there is more time to play with; harder to understand in practice, however, because the much wider range of physical and chemical conditions available in the Galaxy as a whole makes it hard to know where to begin in developing a detailed theory for the emergence of life.

ties. Science fiction writers have long exploited this fascination, which many scientists share. But what are the facts?

On Earth, intelligence seems to have good survival value, and has apparently arisen automatically as a result of evolutionary pressures. Intelligence is found not only in human beings but in other, quite dissimilar creatures such as the dolphins. It is easy to be persuaded that once life has arisen on a planet, it will gradually and systematically evolve into more complex varieties, so that as the competition becomes fiercer, intelligent behavior will gain a selective advantage. Indeed, the jump from microbe to human being seems more readily comprehensible than the jump from prebiotic soup to DNA. According to this philosophy, if life is widespread throughout the Universe, then so are intelligence and, presumably, technology. It is a conclusion that opens the way to an entirely new possibility for discovering extraterrestrial life. Instead of looking for the life-forms themselves, we can look for signs of their technology.

People with poor eyesight could be convinced of the existence of tiny (and possibly intelligent) life-forms on Earth by observing the construction of anthills, without ever seeing, or communicating with, an ant. A hundred years ago, the astronomer Percival Lowell was certain that an advanced civilization had constructed an elaborate network of canals on Mars. Alas, the vague shapes he thought he had perceived through his telescope turned out to owe more to psychology than to physical reality, but the principle of using telescopes to look for technological artifacts on other planets is still sound.

How might a more remote alien community betray its presence to us? The nearest star (after the Sun) is more than 4 light-years (about 24 million million miles) away. Even optimistic estimates do not rate highly the chances of an alien civilization within 10, or even 100, light-years of Earth. Direct observation with optical telescopes over such vast distances is out of the question.

A more promising strategy is to search for radio signals. Radio telescopes are potentially more efficient than their optical counterparts, partly because of the way they can be used in combination, thus multiplying the effective "listening power." Some such systems mimic, at least partially, the properties of a single radio antenna as big as the Earth. Unfortunately, no instrument on Earth is sensitive enough to eavesdrop on the equivalent of our domestic radio and television signals over interstellar distances. But that is largely because such domestic signals spread out in all directions to fill an expanding sphere of space around the source planet (such as the Earth). The situation is changed if powerful radio signals are deliberately directed in a beam toward a certain point in space, when they can have a much longer effective range. The Arecibo radio telescope in Puerto Rico has the power to communicate in this way with a similar device anywhere in our Galaxy—if only we know which stars we should be beaming our messages toward, or listening for messages from.

So existing terrestrial technology is adequate to establish communication with any comparably advanced civilization in our Galaxy. The idea of radio dialogues between advanced communities has caught the imagination of scientists and nonscientists alike, although it is open to many objections. Why should "they" bother to send signals to us? How do "they" even know we exist and possess the technology to detect their signals? And anyway, what is the point of communicating in this way, when even at the speed of light messages are likely to take decades, or more, to reach their destination? Why, also, should "they" use radio anyway, rather than some more advanced technique that we have yet to discover? There could even be some cosmic network of communication between advanced civilizations going on all around us, and we are simply not clever enough to tap in to it.

Searching for ET

Proponents of the communication idea are not dismayed by these problems, for the following reason. The Earth, at 4½ billion years old, is only about one third of the age of the Galaxy, and it has taken about 4 billion years for Earth life to evolve from primitive microorganisms to our modern technological society. If life developed this rapidly on the planets that formed early in the history of the Galaxy, there could have been technological communities well established before the Earth even existed. The capabilities of a technology that has lasted for thousands of years, let alone millions, or even thousands of millions, of years, are incalculable. A little matter like signaling *every* star system in the Galaxy might well be trivial for such an advanced civilization. As for knowing we are here, remember that there *is* that expanding shell of radio and television noise, now more than 50 light-years in radius, spreading outward from the Solar System at the speed of light. A suitably advanced civilization probably could detect this pollution of the cosmic airwaves, even if we could not do so at an equivalent distance. And with millennia upon millennia of history, a transmission time of a few decades might be quite acceptable to such alien intelligences—even if their individual life-spans were as short as our own, which is certainly not something that we should take for granted. Moreover, any alien society with the wit to contemplate establishing contact with a newly arisen technological community (us) would certainly figure out the most probable communication system (radio) that would be appropriate.

If we assume that somebody out there might be trying to get in touch with us, a major obstacle to embarking upon radio communication with aliens is the choice of radio frequency to tune in to. With the whole radio spectrum to choose from, how do we know which portion "they" are transmitting on? Three decades ago, an ingenious suggestion regarding this point was made by Giuseppe Cocconi and Philip Morrison, of MIT. Any community in the Galaxy experienced in the principles of radio telescopes would, they argued, be familiar with the ubiquitous background "hiss" of noise produced by radio emission from the clouds of hydrogen gas strung around the spiral arms of the Milky Way. It is the first thing that any radio astronomer would "hear." What could be more natural, then, than to choose this frequency (or perhaps one half of the frequency, or twice its value, to avoid the noise itself) for interstellar communication? At least, nothing could be more natural if alien thought processes work the same way that those of Cocconi and Morrison do. . . .

Some astronomers have become sufficiently enthusiastic about communicating by radio with aliens that several preliminary efforts have been mounted. Limited searches for incoming signals from nearby star systems have found nothing identifiable as intelligent communication, and a much more ambitious and comprehensive search would be necessary to stand any reasonable chance of success. Undaunted, radio astronomers have sent a burst of transmission from Arecibo toward a giant cluster of stars, far across the Milky Way, where because of the slight spreading of the beam on its journey of tens of millions of light-years the signal will eventually be detectable by any being with a similar radio telescope on any planet orbiting any of the thousands of stars in the cluster. Overall, though, the possibility of success in the search for extraterrestrial intelligence (or SETI, as it is known) is generally regarded as too speculative to command more than a tiny fraction of the available instrument time on the world's major radio telescopes, let alone the construction of the vast array of radio telescopes proposed by some researchers as the minimum requirement for a truly systematic effort.

Where are they?

One of the more sobering conclusions that can be drawn from a fairly simple analysis of the likelihood of extraterrestrial communities concerns the number of other civilizations in our Galaxy that may have already achieved technology. Stars and planets are continually being born in the Galaxy, and as life appears and evolves on each suitable planet, so more and more technological communities might arise. If one is optimistic and assumes that this process is inevitable in any planetary system around a star like the Sun, then the rate at which new communities achieve interstellar radio communication technology is roughly one per decade in the whole Galaxy—one per decade, that is, for at least 10 billion years, if the Galaxy is 14 billion years old and it takes, as it did on Earth, 4 billion years or so for technology to evolve.

This is a doubly remarkable conclusion, for our own radio telescope technology is no more than a few decades old. It follows that we are most probably the newcomers as far as any galactic radio communication club is concerned. *All* the other transmitting communities are likely to be more advanced than us.

The number of such communities around today, however, depends crucially on the life expectancy of a technological civilization, as well as on the birth rate. If Earth civilization is destroyed tomorrow, and if we are typical, that would mean that on average only one community capable of interstellar communication will exist in the Galaxy at any one time. We would hold that distinction today, in lonely isolation, making us the *most* technologically advanced society around in the Galaxy just now. Alternatively, if a typical advanced community survives for, say, 10 million years, then about a million such communities will inhabit the Milky Way at any one time, most of them well ahead of ourselves in technology.

This raises a difficult and intriguing question first posed explicitly in this form by the physicist Enrico Fermi, who among other things was the theorist who gave the neutrino its name. If life is even sporadically extant in the Galaxy at large, then it is hard to see, considering the relative youthfulness of the Earth, how advanced communities would not have arisen millions of years ago. Wouldn't such communities have colonized the entire Galaxy by now?

Consider how this might happen. Imagine our own civilization building a huge spacecraft with an energy supply capable of sustaining life for thousands of years. This could be done with present technology, if the will to do so were there. A few colonists might set out in such a craft at modest speed across the Galaxy in search of a new home. At presently available speeds it may take 10,000 years to reach the nearest star, but eventually some future generation of colonists would establish themselves on another planet. After a further few thousand years that planet would be fully populated, and a further expedition might set out.

Adopting this strategy, the entire Galaxy (which is about 100,000 light-years across) could be populated with humans in just 10 million years—a brief fraction of the age of the Galaxy. In an alternative scenario, would-be Galaxy conquerors could send out robot space probes, only slightly beyond the scope of our present-day technology, equipped with genetic material (frozen samples of egg and sperm, or fertilized eggs, or even the raw materials of living molecules together with genetic information coded into the computer brain of the robot, ready to manufacture DNA on arrival), almost literally to seed any suitable planets with Earth life. And although many people might doubt the likelihood of any civilization being motivated to do this, even if it is technologically feasible, remember that it takes only *one* such colonizing species to arise any time during the Galaxy's roughly 14 billion year lifetime to date,[6] and the Milky Way would be teeming with their descendants by now. So where are they?

The dilemma would seem to be a serious one for those who believe in the existence of intelligent life elsewhere in the Universe. Perhaps they *are* here, but we are too dim to notice—much as ants go about their business oblivious of human scrutiny. Perhaps, as some UFO buffs would have us believe, Earth is being watched from a distance, a kind of cosmic nature reserve on which "no trespassing" signs have been erected, for reasons we cannot comprehend. Or, perhaps, there is an inbuilt self-destruct mechanism in all technological societies aggressive enough to indulge in colonization, which destroys them before they can reach the point of interstellar travel. Maybe the very evolutionary pressures that lead to intelligence also lead to aggression, and at some critical point the combination of the two leads to nuclear annihilation or the equivalent—or the intelligent species inevitably runs wild across its home planet, despoiling the environment and removing the planet's capacity to support life. Slightly less gloomily, it may be that there are problems with interstellar travel we have not thought of; or (improbably) per-

6 One colonizing civilization, that is, out of roughly a billion technological civilizations that emerge, according to the figures we used before.

haps Earth life is of such an exotic type that our planet would be inhospitable to most life-forms. Surely it cannot be that we are the only technological civilization ever to have arisen in the Galaxy, or in the entire Universe?

From matter to mind

In an article entitled "Information, Physics, Quantum: The Search for Links," the theoretical physicist John Wheeler claimed, at the end of the 1980s, that there was no escape from the conclusion that "The world cannot be a giant machine, ruled by any pre-established continuum physical law." It would be more accurate, opined Wheeler, to think of the physical Universe as a gigantic information-processing system in which the output was as yet undetermined. As an emblem of this massive paradigm shift, he coined the slogan "It from bit." That is to say, every *it*—every particle, every field of force, even spacetime itself—is ultimately manifested to us through *bits* of information.

The process of science is a process of interrogation of nature. Each experimental measurement, each observation, elicits answers from nature in terms of information bits. But more fundamentally, the essentially quantum nature of the physical world ensures that, at rock bottom, all such measurements and observations are reduced to answers of the simple "yes-no" kind. Is an atom in its ground state? Yes. Is an electron's spin pointing up? No. And so on. And because of the inherent uncertainty of quantum physics, these answers cannot be foretold. Moreover, as we discussed in Chapter 7, the observer plays a key role in deciding the outcome of the quantum measurements—the answers, and the nature of reality, depend in part on the questions asked.

Wheeler is an extreme exponent of the "participatory universe" philosophy, in which observers are central to the nature of physical reality, and matter is ultimately relegated to mind. Another proponent of these ideas is Frank Tipler, of Tulane University in New Orleans. Tipler's position differs from Wheeler's, however, in suggesting that the observer participation in nature is as yet only trivial. Instead, Tipler believes that intelligence will eventually spread throughout the cosmos, participating more and more in the workings of nature, until it eventually reaches such an extent that it will have *become* nature. According to Tipler, intelligent life—or more likely a network of computing devices—will spread out from its planet of origin (possibly Earth), and slowly but surely gain control over larger and larger domains. Tipler envisages not just the Solar System, nor even the Galaxy alone, but the entire Universe coming under control of this manipulative intelligence—a scenario that echoes in some ways the earlier philosophical speculations of the Jesuit Pierre Teilhard de Chardin, but in which technology is the key. Although the process may take trillions of years, in Tipler's scenario the upshot of this creeping "technologization" of nature will be the amalgamation of the whole cosmos into a single intelligent computing system! In effect, intelligence will have hijacked the "natural" information-processing system we call the Universe, and used it for its own ends. All the "its" will be turned back into "bits."

We mention these admittedly speculative ideas to illustrate the profound change in perspective that has accompanied the move toward a postmechanistic paradigm. In place of clodlike particles of matter in a lumbering Newtonian machine we have an interlocking network of information exchange—a holistic, indeterministic and open system—vibrant with potentialities and bestowed with infinite richness. The human mind is a by-product of this vast informational process, a by-product with the curious capability of being able to *understand*, at least in part, the principles on which the process runs.

Descartes founded the image of the human mind as a sort of nebulous substance that exists independently of the body. Much later, in the 1930s, Gilbert Ryle derided this dualism in a pithy reference to the mind part as "the ghost in the machine." Ryle articulated his criticism during the triumphal phase of materialism and mechanism. The "machine" he referred to was the human body and the human brain, themselves just parts of the larger cosmic machine. But already, when he coined that pithy expression, the new physics was at work, undermining the world view on which Ryle's philosophy was based. Today, on the brink of the twenty-first century, we can see that Ryle was right to dismiss the notion of the ghost in the machine—not because there is no ghost, but because there is no machine.

Bibliography

The book you now hold represents a synthesis of ideas, reflecting a growing awareness of the need for a new paradigm of wholeness within the physical sciences that has grown up in each of the authors as a result of investigating and writing about a wide variety of new ideas in physics in recent years. Like the Universe itself, the whole story, as related in this book, is greater than the sum of its parts, which we (among others) have discussed before. The following books amplify some of the topics that make up our own synthesis. We include many of our own earlier books in the list because these reflect the growth of our own dissatisfaction with the purely reductionist approach, and provide the detailed background for what is (we hope) a convincing overview in the present book.

Barrow, John, *The World within the World* (Oxford University Press, 1988).

Barrow, John, *Theories of Everything* (Oxford University Press, 1991).

Barrow, John, and Silk, Joseph, *The Left Hand of Creation* (New York: Basic Books, 1983).

Barrow, John, and Tipler, Frank, *The Anthropic Cosmological Principle* (Oxford University Press, 1986).

Berkeley, George, "On the Principles of Human Knowledge," in *The Works of George Berkeley,* ed. Alexander Campbell Fraser (Oxford: Clarendon Press, 1901).

Bohm, David, *Wholeness and the Implicate Order* (London: Routledge & Kegan Paul, 1980).

Clark, David, *Superstars* (London: Dent, 1979).

Clayton, Donald, *The Dark Night Sky* (New York: Quadrangle/New York Times, 1975).

Coveney, Peter, and Highfield, Roger, *The Arrow of Time* (London: W. H. Allen, 1990).

Davies, Paul, *The Physics of Time Asymmetry* (San Francisco: University of California Press, 1974).

Davies, Paul, *The Accidental Universe* (Cambridge University Press, 1982).

Davies, Paul, *Superforce* (New York: Simon & Schuster, 1984).

Davies, Paul, *The Cosmic Blueprint* (New York: Simon & Schuster, 1987).

Davies, Paul, *The Mind of God* (New York: Simon & Schuster, 1991).

Davies, Paul (ed.), *The New Physics* (Cambridge University Press, 1989).

Davies, Paul, and Brown, Julian, *The Ghost in the Atom* (Cambridge University Press, 1986).

Dunne, J. W., *An Experiment with Time* (London: Faber, 1934).

Feinberg, Gerald, and Shapiro, Robert, *Life beyond Earth* (New York: Morrow, 1980).

Ferris, Timothy, *Coming of Age in the Milky Way* (New York: Morrow, 1988).

Feynman, Richard, *The Character of Physical Law* (London: BBC Books, 1966).

Feynman, Richard, *QED: The Strange Theory of Light and Matter* (Princeton, NJ: Princeton University Press, 1985).

Gilder, George, *Microcosm: The Quantum Revolution in Economics and Technology* (New York: Simon & Schuster, 1989).

Gleick, James, *Chaos* (New York: Simon & Schuster, 1987).

Gribbin, John, *In Search of Schrödinger's Cat* (New York: Bantam, and London: Black Swan, 1984).

Gribbin, John, *In Search of the Double Helix* (New York: Bantam, and London: Black Swan, 1985).

Gribbin, John, *In Search of the Big Bang* (New York: Bantam, and London: Black Swan, 1986).

Gribbin, John, *The Omega Point* (New York: Bantam, and London: Black Swan, 1987).

Gribbin, John, *Hothouse Earth: The Greenhouse Effect and Gaia* (New York: Grove, and London: Black Swan, 1990).

Gribbin, John, *Blinded by the Light: The Secret Life of the Sun* (New York: Harmony, and London: Black Swan, 1991).

Gribbin, John, and Rees, Martin, *Cosmic Coincidences* (New York: Bantam, and London: Black Swan, 1989).

Harrison, Ed, *Cosmology* (Cambridge University Press, 1981).

Hawking, Stephen, *A Brief History of Time* (New York: Bantam, 1988).

Heisenberg, Werner, *Physics and Philosophy* (London: Penguin, 1989).

Hofstadter, Douglas, *Gödel, Escher, Bach* (London: Penguin, 1979).

Kaufmann, William J., *The Cosmic Frontiers of General Relativity* (Boston: Little, Brown, 1977).

Lovelock, James, *Gaia* (Oxford University Press, 1979).

Lovelock, James, *The Ages of Gaia* (Oxford University Press, 1988).

Mandelbrot, Benoit, *The Fractal Geometry of Nature* (San Francisco: W. H. Freeman, 1982).

Minsky, Marvin, *The Society of Mind* (New York: Simon & Schuster, 1987).

Newton, Isaac, *Principia Mathematica* (1687); available in English as *Mathematical Principles of Natural Philosophy*, in the series Great Books of the Western World (Chicago: Encyclopedia Britannica, 1952).

Pagels, Heinz, *The Dreams of Reason* (New York: Simon & Schuster, 1988).

Penrose, Roger, *The Emperor's New Mind* (Oxford University Press, 1989).

Prigogine, Ilya, and Stengers, Isabelle, *Order Out of Chaos* (London: Heinemann, 1984).

Rae, Alastair, *Quantum Physics: Illusion or Reality?* (Cambridge University Press, 1986).

Rafelski, Johann, and Muller, Berndt, *The Structured Vacuum: Thinking about Nothing* (London: Deutsch, 1985).

Shipman, Harry, *Black Holes, Quasars and the Universe* (New York: Houghton Mifflin, 1976).

Shklovskii, Iosif, and Sagan, Carl, *Intelligent Life in the Universe* (San Francisco: Holden-Day, 1966).

Stewart, Ian, *Does God Play Dice?* (Oxford: Blackwell, 1989).

Weinberg, Steven, *The First Three Minutes* (Glasgow: Collins, 1977).

Will, Clifford, *Was Einstein Right?* (New York: Basic Books, 1986).

More general information about astronomy, and reviews of new books on astronomy and astrophysics, can be found in the pages of *Mercury,* the journal of the Astronomical Society of the Pacific (390 Ashton Avenue, San Francisco, CA 94112, USA).

Index

Printed in the United States
By Bookmasters